Technology, Peace and Security I
Technologie, Frieden und Sicherheit

Series Editor

Christian Reuter, Darmstadt, Germany

Technology plays a crucial role in various aspects of peace and security. This book series explores the intersection of computer science with peace and security studies. With a focus on cyber security and privacy, human-computer interaction as well as peace and conflict studies, it addresses topics such as peace informatics and technical peace research (cyber war, peace, arms control, and dual use), crisis informatics and information warfare (social media and collaborative technologies in conflict and crisis situations, misinformation and opinion manipulation), as well as usable safety, security, and privacy (resilient digital infrastructures, security and privacy enhancing technologies).

Technologie spielt eine entscheidende Rolle in verschiedenen Aspekten von Frieden und Sicherheit. Diese Buchreihe befasst sich mit Fragen an der Schnittstelle der Informatik mit der Friedens- und Sicherheitsforschung. Mit einem Fokus auf Cybersicherheit und Privatheit, Mensch-Computer-Interaktion sowie Friedens- und Konfliktforschung werden Themen wie Friedensinformatik und technische Friedensforschung (Cyberkrieg, Frieden, Rüstungskontrolle und Dual-Use), Kriseninformatik und Informationskrieg (soziale Medien und kollaborative Technologien in Konflikt- und Krisensituationen, Desinformation und Meinungsmanipulation) sowie nutzbare Sicherheit und Privatheit (resiliente digitale Infrastrukturen, Technologien zur Verbesserung der Sicherheit und des Datenschutzes) behandelt.

Sebastian Linsner

Privacy Preserving Data Management

Assisting Users in Data Disclosure Scenarios

Springer Vieweg

Sebastian Linsner
Informatik
Technische Universität Darmstadt
Darmstadt, Germany

Dissertation approved by the Department of Computer Science of the Technical University of Darmstadt for the award of the academic degree Doktor-Ingenieur (Dr.-Ing.)
by Sebastian Linsner, M. Sc.
First examiner: Prof. Dr. Dr. Christian Reuter
Second examiner: Prof. Anna Spagnolli
Date of submission: 24.06.2024
Date of disputation: 05.08.2024

ISSN 3004-9318 ISSN 3004-9326 (electronic)
Technology, Peace and Security I Technologie, Frieden und Sicherheit
ISBN 978-3-658-47444-7 ISBN 978-3-658-47445-4 (eBook)
https://doi.org/10.1007/978-3-658-47445-4

This Springer Vieweg imprint is published by the registered company Springer Fachmedien Wiesbaden GmbH, part of Springer Nature.
The registered company address is: Abraham-Lincoln-Str. 46, 65189 Wiesbaden, Germany

Foreword

Users often need to share their data but are not always informed about its use or consequences. The PhD thesis of Sebastian Linsner aims to help users by incorporating their perspectives into the development of digital tools, focusing on two main scenarios: everyday private use and business-to-business (B2B) collaborations, particularly for small and medium enterprises (SMEs) in agriculture. SMEs in this sector need to share data for work processes, making it crucial to understand their needs for data sovereignty. Sebastian's research addresses three core aspects: identifying problems and risks of data disclosure, finding ways developers can mitigate these risks, and ensuring the practical adoption of these solutions. For the B2B sector, the findings highlight the neglect of smaller businesses' privacy needs, slowing digital transformation. Farmers expressed a strong need for tools to protect their business strategies. Transparency was shown to build trust and correct false security perceptions. Customizable data flow controls were also investigated, revealing that security features can enhance privacy and usability without major trade-offs. Overall, the thesis emphasizes the importance of enabling users to make informed decisions to maintain data sovereignty, which makes an important contribution.

Prior to starting his PhD, Sebastian attended the first lecture after the foundation of PEASEC, "Safety critical human-computer interaction", and then successfully applied for a research associate position. He first worked in our new BMBFfunded project *HyServ* (hybrid services in agriculture), and later he deepened his focus on privacy in the DFG Collaborative Research Center CROSSING (transparency as a user-centred intervention for privacy and security) after being heavily involved in our sub-project application. The studies included in his resulting PhD thesis have been published as six peer-reviewed papers. In total, he has been involved in 17 publications. Additionally, he has been actively involved in the lecture "General Computer Science 2".

Sebastian Linsner has demonstrated his ability to conduct independent scientific research. Consequently, his dissertation was accepted by the Department of Computer Science at the Technical University of Darmstadt in August 2024 for the degree of *Dr.-Ing.*, marking the sixth PhD thesis in our research group PEASEC. Sebastian, thank you for your contributions and for allowing me to accompany you on your journey to your PhD. I wish you all the best and every success in the future.

Prof. Dr. Dr. Christian Reuter
Professor for Science and Technology
for Peace and Security (PEASEC)
and Dean of the Department
of Computer Science
at Technical University of Darmstadt
Darmstadt, Germany

Acknowledgements

First of all, I want to express my gratitude to those who contributed to this dissertation. This thesis was only possible because many people supported me throughout my time as a doctoral researcher. Firstly, I want to thank my doctoral supervisor, *Prof. Dr. Dr. Christian Reuter*, for his continuous support in all of my research endeavors. Without his guidance in designing research approaches and the handling of scientific processes, the underlying publications of this thesis would not have been possible. I want to thank *Prof. Dr. Anna Spagnolli* for being my co-examiner and *Prof. Dr. Marc Fischlin* and *Prof. Dr.-Ing. Jörn Kohlhammer* for joining the doctoral committee.

Likewise, I want to thank my colleagues and the entire team at PEASEC for providing me with a friendly, welcoming, and productive work environment. The time we spent together was fun and filled with interesting conversations. In particular, I want to thank *Dr.-Ing. Franz Kuntke* for the many pleasant years we shared an office and worked on joint research projects. Furthermore, I want to thank my other co-authors *Tom Biselli*, *Prof. Dr.-Ing. Lucas Vincenzo Davi*, *Kilian Demuth*, *Jonas Franken*, *Dr. Marc-André Kaufhold*, *Gina Maria Schmidbauer-Wolf*, *Enno Steinbrink*, and *Sebastian Surminski* for their diverse expertise and the perspectives they provided to enhance my research.

I want to thank all the others who made the scientific results of this thesis possible: student assistants who supported me with proof-reading, formatting texts, or contributions to demonstrators, especially *Tobias Frey* and *Laura Utz* for the implementation of the demonstrator used in Paper C and *Paula Meißner* for proof-reading this thesis, as well as the numerous anonymous researchers that peer-reviewed my papers to improve the quality and ensured the scientific contribution.

Furthermore, a thank you goes to the projects which funded me during my time as a PhD student and allowed me to conduct interesting research: *CROSSING* (funded by the Deutsche Forschungsgemeinschaft DFG–SFB 1119 – 236615297), *HyServ* (funded by the German Federal Ministry for Education and Research (BMBF) (01IS17030B)), and *PriVis* (funded by BMBF and the Hessian Ministry of Higher Education, Research, Science and the Arts within their joint support of the National Research Center for Applied Cybersecurity ATHENE). The people who were brought together by the projects were a huge inspiration for me and provided me with valuable feedback.

Since this thesis features many qualitative methods, I am very grateful to all those who offered their time and expertise to take part in the interviews and evaluations. Without their commitment, the findings of this thesis would not exist.

Finally, I want to thank my family, who was a steady support for me during my time as a doctoral researcher and who always backed me up on this entire journey.

Abstract

Users face many situations in which they have to disclose their data to others. In many cases, they are not properly informed about the usage of their data or the consequences of data disclosure. This thesis aims to support users in these situations by investigating possibilities to incorporate their perspectives into the development process for digital tools. The investigated scenarios are twofold: Firstly, business-to-business (B2B) collaborations, especially highlighting the perspective of owners of small and medium enterprises (SMEs) who are forced to provide data for business processes. This holds especially true for the domain of agriculture, where many work processes are coordinated with a division of work, requiring data exchange. Furthermore, SMEs are predominant in German agriculture, making this domain ideal for the investigation. Secondly, scenarios of everyday usage for private users that require the disclosure of data are investigated. These revolve around browser tracking, attestation of internet of things (IoT) devices to prevent espionage through malware and a novel interaction concept for increased privacy in customer-to-business (C2B) interactions. Although both scenarios have some similarities, considering them separately broadens the perspective on software development. Extending personally owned data to operational data especially raises new demands for development processes that preserve data sovereignty. In contrast to private users, in B2B collaborations, declining the usage of digital tools and data disclosure is not possible because business owners have to adhere to regulations and provide business partners with data for collaborative work processes.

This thesis investigates how users can be supported in these scenarios by researching three core aspects: Firstly, problems and risks that lead to (unintended) data disclosure or arise from it have to be identified. Secondly, measures are investigated that developers can apply to mitigate these risks and meet the needs of the users. Finally, the developed solutions have to be adopted into practice. Strategies to bring security- and privacy-enhancing technologies to use are essential because a tool that is not used cannot protect anyone. These three research directions are investigated with several publications that form the main body of this cumulative thesis. Using their findings, the over-arching research question is answered: **"How can users be enabled to make informed decisions to maintain data sovereignty and manage their data in data disclosure scenarios?"**

The results for the B2B sector show a significant overlook of perspectives of smaller businesses, neglecting their privacy needs and thus slowing down the digital transformation. The underlying publications identify several technical, social and ethical aspects for technology adoption in this domain. Farmers who participated in these studies had an especially strong demand for tools that allow them to enact their data sovereignty and prevent others from getting insights into their business strategy.

For private users, the contributions are threefold: Firstly, different levels of support for users were investigated, which led to the creation of support personas, a novel approach representing different needs and preferences regarding the level of support users need to interact with security- and privacy-enhancing tools. Secondly, transparency was evaluated as a suitable way to educate users and build trust in security mechanisms. The evaluation showed that false mental models were revoked by the increased understanding, allowing users to estimate the level of protection correctly and avoiding data leakage through overestimation of protection. Finally, different options to control the data flow in everyday scenarios were investigated. By providing different levels of customization, light was shed on the effects of the control paradox. Furthermore, the evaluation showed that incorporating security and privacy features into everyday use cases does not require a severe trade-off regarding usability. Instead, the proposed interaction concept was perceived as an overall improvement to privacy and usability.

To conclude, this thesis presents different approaches to *identify* risks and problems in data disclosure scenarios for private users as well as business owners, explores solutions for *developers* and investigates strategies for the *adoption* of security- and privacy-enhancing technologies into practice.

Author's Publications

In sum, 17 publications have been published in the context of the authors work. The following 6 publications are published as chapters in part II of this thesis:

1. S. Linsner, F. Kuntke, E. Steinbrink, J. Franken, and C. Reuter, "The Role of Privacy in Digitalization – Analysing the German Farmers' Perspective," *Proceedings on Privacy Enhancing Technologies (PoPETs)*, vol. 2021, no. 3, pp. 334–350, 2021. https://doi.org/10.2478/popets-2021-0050
2. S. Linsner, F. Kuntke, G. M. Schmidbauer-Wolf, and C. Reuter, "Blockchain in der Landwirtschaft 4.0 – Empirische Studie zu Erwartungen von Landwirt innen gegen über dezentralen Services auf Basis von Distributed Ledger Technology," in *Mensch und Computer – Tagungsband*, F. Alt, A. Bulling, and T. Döring, Eds., Hamburg, Germany: ACM, 2019, pp. 103–113. https://doi.org/10.1145/3340764.3340799
3. S. Linsner, E. Steinbrink, F. Kuntke, J. Franken, and C. Reuter, "Supporting Users in Data Disclosure Scenarios in Agriculture through Transparency," *Behaviour & Information Technology (BIT)*, vol. 41, no. 10, pp. 2137–2159, 2022. https://doi.org/10.1080/0144929X.2022.2068070
4. K. Demuth, S. Linsner, T. Biselli, M.-A. Kaufhold, and C. Reuter, "Support Personas: A Concept for Tailored Support of Users of Privacy-Enhancing Technologies," *Proceedings on Privacy Enhancing Technologies (PoPETs)*, no. 4, pp. 797–817, 2024, ISSN: 2299-0984. https://doi.org/10.56553/popets-2024-0142. [Online]. Available: https://petsymposium.org/popets/2024/popets-2024-0142.pdf

5. S. Linsner, K. Demuth, S. Surminski, L. Davi, and C. Reuter, "Building Trust in Remote Attestation Through Transparency – A Qualitative User Study on Observable Attestation," *Behaviour & Information Technology (BIT)*, pp. 1–21, 2024, Publisher: Taylor & Francis, ISSN: 0144-929X. https://doi.org/10.1080/0144929X.2024.2374889. [Online]. Available: https://doi.org/10.1080/0144929X.2024.2374889
6. S. Linsner, K. Demuth, M. Fischlin, and C. Reuter, "Decision-based Data Distribution (D^3): Enabling Users to Minimize Data Propagation in Privacysensitive Scenarios," *Proceedings on Privacy Enhancing Technologies (PoPETs)*, no. 4, 2024, ISSN: 2299-0984. https://doi.org/10.56553/popets-2024-0113. [Online]. Available: https://petsymposium.org/popets/2024/popets-2024-0113.php

The following 11 papers are not included in the thesis, although their findings are supplementary to it:

7. G. R. Chandran, K. Demuth, K. Edalatnejad, S. Linsner, C. Reuter, and T. Schneider, "Encrypted MultiChannel Communication (EMC2): Johnny should use secret sharing," in 23. *Workshop on Privacy in the Electronic Society (WPES'24)*, Salt Lake City, USA: ACM, 2024, pp. 34–39. https://doi.org/10.1145/3689943.3695051. [Online]. Available: https://dl.acm.org/doi/10.1145/3689943.3695051
8. F. Kuntke, M.-A. Kaufhold, S. Linsner, and C. Reuter, "GeoBox: Design and Evaluation of a Tool for Resilient and Decentralized Data Management in Agriculture," *Behaviour & Information Technology (BIT)*, vol. 43, no. 4, pp. 764–786, 2024, Publisher: Taylor & Francis, ISSN: 0144-929X. https://doi.org/10.1080/0144929X.2023.2185747
9. S. Surminski, C. Niesler, S. Linsner, L. Davi, and C. Reuter, "SCAtt-man: Side-Channel-Based Remote Attestation for Embedded Devices that Users Understand," in *Proceedings of the Thirteenth ACM Conference on Data and Application Security and Privacy*, ser. CODASPY '23, New York, NY, USA: Association for Computing Machinery, 2023, pp. 225–236, ISBN: 9798400700675. https://doi.org/10.1145/3577923.3583652
10. E. Steinbrink, T. Biselli, S. Linsner, F. Herbert, and C. Reuter, "Privacy Perception and Behaviour in Safety-Critical Environments," in *Human Factors in Privacy Research*, N. Gerber, A. Stöver, and K. Marky, Eds., Cham: Springer International Publishing, 2023, pp. 237–251, ISBN: 978-3-031-28643-8. [Online]. Available: https://doi.org/10.1007/978-3-031-28643-8_12

11. F. Kuntke, S. Linsner, E. Steinbrink, J. Franken, and C. Reuter, "Resilience in Agriculture: Communication and Energy Infrastructure Dependencies of German Farmers," *International Journal of Disaster Risk Science (IJDRS)*, vol. 13, no. 2, pp. 214–229, 2022, ISSN: 2192-6395. https://doi.org/10.1007/s13753-022-00404-7
12. L. Buhleier, S. Linsner, E. Steinbrink, and C. Reuter, "Eine Klassifikation sicherheitskritischer UX-Design-Patterns," de, in *Mensch und Computer – Workshopband*, ser. Mensch und Computer 2022 -Workshopband, Darmstadt: Gesellschaft für Informatik, 2022. https://doi.org/10.18420/muc2022-mciws10-275
13. F. Kuntke, V. Romanenko, S. Linsner, E. Steinbrink, and C. Reuter, "Lo-RaWAN Security Issues and Mitigation Options by the Example of Agricultural IoT Scenarios," *Transactions on Emerging Telecommunications Technologies (ETT)*, vol. 33, no. 5, e4452, 2022, ISSN: 2161-3915. https://doi.org/10.1002/ett.4452
14. F. Kuntke, M. Sinn, S. Linsner, and C. Reuter, "Low PowerWide Area Networks (LPWAN) für krisentaugliche Datenübertragung in landwirtschaftlichen Betrieben," in *41. GIL-Jahrestagung: Informatik in der Land-, Forst- und Ernährungswirtschaft*, A. Meyer-Aurich, M. Gandorfer, C. Hoffmann, C. Weltzien, S. D. Bellingrath-Kimura, and H. Floto, Eds., Meyer-Aurich, Andreas Gandorfer, Markus Hoffmann, ChristaWeltzien, Cornelia Bellingrath-Kimura, Sonoko D. Floto, Helga: Gesellschaft für Informatik, 2021, pp. 193–198. [Online]. Available: http://www.peasec.de/paper/2021/2021_KuntkeSinnLinsnerReuter_LowPowerWideAreaNetworks_GIL.pdf
15. A. Bernardi, S. Becker, C. Struve, S. Linsner, C. Reuter, and G. M¨ uller, "Erfolgsorientierte Dienstleistung: Neue Perspektiven für die landwirtschaftliche Arbeitsteilung durch hybride Dienstleistungen," in *41. GIL-Jahrestagung: Informatik in der Land-, Forst- und Ernährungswirtschaft*, A. Meyer-Aurich, M. Gandorfer, C. Hoffmann, C.Weltzien, S. D. Bellingrath-Kimura, and H. Floto, Eds., Gesellschaft für Informatik, 2021, p. 37. [Online]. Available: https://peasec.de/paper/2021/2021_Bernardietal_ErfolgsorientierteDienstleistung_GIL.pdf
16. S. Linsner, R. Varma, and C. Reuter, "Vulnerability Assessment in the Smart Farming Infrastructure through Cyberattacks," in *39. GIL-Jahrestagung: Informatik in der Land-, Forst- und Ernährungswirtschaft Fokus; Digitalisierung für landwirtschaftliche Betriebe in kleinstrukturierten Regionen – ein Widerspruch in sich?, Lecture Notes in Informatics (LNI)*, A. Meyer-Aurich, Ed.,Wien, Austria: Gesellschaft für Informatik, 2019, pp. 119–124. [Online]. Available: http://gil-net.de/Publikationen/139_119.pdf

17. A. Bernardi, C. Reuter, W. Schneider, S. Linsner, and M.-A. Kaufhold, "Hybride Dienstleistungen in digitalisierten Kooperationen in der Landwirtschaft," in *39. GIL-Jahrestagung: Informatik in der Land-, Forst- und Ernährungswirtschaft Fokus; Digitalisierung für landwirtschaftliche Betriebe in kleinstrukturierten Regionen – ein Widerspruch in sich?, Lecture Notes in Informatics (LNI)*, A. Meyer-Aurich, Ed., Vienna, Austria: Gesellschaft fürInformatik, 2019, pp. 25–30. [Online]. Available: http://gil-net.de/Publikationen/139_25-30.pdf

Contents

Part I Synopsis

Part II Publications

List of Figures

List of Tables

Part I
Synopsis

Introduction

1

1.1 Motivation and Problem Statement

Users face countless situations in which their data is disclosed. Once this data is made available to others, users lose their sovereignty over it. Third parties can access, process, and use the data for their own purposes. This could have negative consequences for the original data owners. In this work, we use two different scenarios to demonstrate possible risks originating from the loss of data sovereignty and present approaches how to mitigate those risks. The first scenario involves business collaboration and the sharing of operational data. To demonstrate how data can be (mis-)used, the domain of digitalized agriculture is investigated. In this domain, the development of digital services is mostly driven by big companies with major market power. On the other hand, there is a multitude of farmers who own small and medium enterprises, which, in many cases, are family-run [18]. These small actors in the supply chain face serious risks by losing sovereignty over their data. In this context, sovereignty means the ability to have full control over their data [19]. Those risks range from personalized prices to cut their profit to buying up smaller farms and using the collected data for large-scale operations with a cheaper workforce [20]. These risks do not only form a serious barrier to the adoption of new technologies, they also require measures to raise awareness of the consequences of data disclosure.

The second scenario involves everyday online interaction. End users face many situations where third parties want to acquire their data for different purposes. By choosing user-unfriendly designs, users are nudged towards poor privacy decisions [12]. Popular methods are interfaces that make opting out of data collection difficult and psychological strategies that trick users into clicking the wrong button, thus gaining their consent for data collection. In everyday situations, users access the

S. Linsner, *Privacy Preserving Data Management*, Technology, Peace and Security I Technologie, Frieden und Sicherheit,
https://doi.org/10.1007/978-3-658-47445-4_1

internet for different purposes and, ideally, to achieve their tasks efficiently. One example where this is already researched are so-called cookie-banners. Cookie notices which require a multitude of clicks to opt out of data collection and tracking pose a hindrance to that goal. If the users get annoyed by those unfriendly notices, they tend to accept tracking in order to resume their original task. Furthermore, users usually think of themselves as a "good" person. This leads to a bias known as social desirability [21] which can occur in interview situations: Users answer in a way that they think is socially accepted to convince others that they are a "good" person. This is exploited in online scenarios by implying that denying the tracking and data collection by third parties is socially undesirable. These patterns try to make the user feel bad by denying tracking and disturbing the self-image of a "good" person. Users have to be educated in order to recognize and comprehend those situations in their online interactions. One way of achieving this is by making the mechanisms in the background transparent for users. If they can inform themselves about the functionality in the situation they are facing a security- or privacy-related mechanism without researching the topic themselves, they can be assisted in making better decisions. The correct assessment of security and privacy is a crucial aspect of online interactions: Users need to know what a mechanism actually does in order to use it with ease of mind and prevent overestimation of security. For example, if users simply estimate that they are protected from every harm by installing an anti-virus program, they might be too careless and might risk data leakage.

Instead of educating the users to make better decisions themselves, developers could simply eliminate the factor of poor privacy decisions and enforce security and privacy by design. However, this approach has proven to be lacking: If users have the decisions taken away from them, they feel paternalized or helpless, being at the mercy of the company providing them with tools. Users have to trust companies that the decisions they make for them are actually good and that companies live up to their claims and promises. One popular example is messengers claiming to have end-to-end encryption while there are rumors that governmental authorities still can read all messages [22]. This leads to problems such as trying to circumvent security mechanisms or discarding the tool altogether. This phenomenon is called reactance [23]. And if security mechanisms or privacy-enhancing technologies are not used, they cannot protect anyone. Therefore, this thesis incorporates users into the development process and provides transparency by communicating how solutions work.

The two different perspectives (B2B relation and privacy users) are used to broaden the perspective of this thesis. While many problems are similar, each domain has unique issues that affect the development process of digital tools. Whether they have to share operational data in a business context or just use their private browser to surf the internet. In each case, third parties want to acquire data from the users. Often, the terms are not favorable for users: Either controls to configure the data disclosure are designed to be user unfriendly to nudge the users towards poor privacy decisions, or users are required to upload their data to third-party servers for processing and sharing. The adoption of technology differs especially: In most private user scenarios, users can simply choose not to use a tool if they do not like the privacy policies of the manufacturer. This choice is affected by different social factors such as peer pressure [24] or Fear-Of-Missing-Out (FOMO) [25], but ultimately, the users still have a choice. On the other hand, B2B relations sometimes enforce the adoption of tools upon individuals since data exchange is essential for processes alongside the supply chain, especially in collaborative work processes. Therefore, expanding the scope of this thesis allows for a broader perspective on the needs and preferences that affect the development of software solutions.

1.2 Aim and Research Question

This work investigates how users can be supported to understand security and privacy mechanisms better and gain more control over their data. As a first step, this work investigates users' behavior and preferences to determine where and how users can be supported in data disclosure scenarios. Secondly, this work presents prototypes that were developed to address the problems and challenges of these scenarios, which will be refined by user testing and usability evaluations. Thirdly, we investigate the adoption of our prototypes into practice, as hindrances to adoption have to be identified and implications of using new technologies need to be explored. In this way, this work derives challenges for user assistance in data disclosure scenarios and provides solutions to address these.

Hence, this thesis tries to answer the following main research question:

RQ **How can users be supported in data disclosure scenarios to make informed decisions and manage their data in a privacy-preserving manner?**

In order to answer the main research question, three different areas are investigated. This way, different aspects can be highlighted, and their role in data disclosure scenarios can be researched. Therefore, three sub-questions are formulated, representing one particular research focus of this thesis:

RQ1 **Which problems and risks can arise from or lead to (unintentional) data disclosure in B2B relations and private everyday usage?**

First, before users can be assisted in mitigating risks, those risks have to be identified. For this purpose, the scenarios of data sharing in business collaborations and tracking in everyday online interactions are investigated. For a better overview, all content related to answering the first sub-question is referred to as the research focus IDENTIFY.

RQ2 **How can developers design tools to mitigate risks in data disclosure scenarios?**

Secondly, in order to support users in the right way, different approaches have to be investigated to develop new tools. Users are a heterogeneous group and not every approach is suited for every user. To find the best forms of assistance for each user, different approaches are investigated. All solutions and recommendations for developers to provide users with tools to protect their data or enact their data sovereignty are referred to as research focus DEVELOP.

RQ3 **How can users be encouraged to adopt technology in order to protect their own data?**

Finally, after identifying possible risks and suitable ways to assist the users, the adoption of the technology has to be investigated. A security or privacy mechanism that is not used cannot protect anyone. Strategies to bring those technologies into practice are investigated in the research focus ADOPT.

1.3 Content and Structure of the Thesis

This dissertation consists of two parts: a synopsis (I) and publications (II).
Part I: Synopsis

Chapter 1 (*Introduction*) presents the motivation for this thesis and derives the main research question and the three sub-questions. Furthermore, the underlying publications are introduced alongside the authors' statements which point out the contribution of each individual author to the joint publication.

Chapter 2 (*Research Design*) gives an overview of the applied methods and their role in each publication.

Chapter 3 (*Findings*) accumulates the findings from the individual publications that are relevant to each research focus (IDENTIFY, DEVELOP, and ADOPT).

Chapter 4 (*Discussion*) refines the findings to form an overarching model that is used to answer the research questions. Furthermore, the limitations are discussed and future work is outlined.

Chapter 5 (*Conclusion*) sums up the core contributions of this thesis.

Part II: Publications

The second part of the dissertation consists of the publications that provide the empirical and theoretical findings to answer the research questions. They have been previously published as articles or conference papers. Major parts of this dissertation are written by several authors. The purpose of this section is to list the independent academic contributions that have been approved by all authors.

Part II consists of two segments: The first segment comprises publications concerned with data sharing in the business context. The second segment comprises the publications revolving around private users.

Chapter 6 (*Paper A: The Role of Privacy in Digitalization—Analyzing Perspectives of German Farmers*) Technological progress can disrupt domains and change the way we work and collaborate. This paper presents a qualitative study with 52 German farmers that investigates the impact of the ongoing digitalization process in agriculture and discusses the implications for privacy research. As in other domains, the introduction of digital tools and services leads to the data itself becoming a resource. Sharing this data with products along the supply chain is favored by retailers and consumers, who benefit from traceability through transparency. However, transparency can pose a privacy risk. Having insight into the business data of others along the supply chain provides an advantage in terms of market position. This is particularly true in agriculture, where there is already a significant imbalance of power between actors. A multitude of small and medium-sized farming businesses are opposed by large upstream and downstream players that drive technological innovation. Further weakening the market position of farmers could lead to severe consequences for the entire sector. We found that, on the one hand, privacy behavior are affected by adoption of digitalization, and on the other hand, privacy itself influences adoption of digital tools. Our study sheds light on the emerging challenges for farmers and the role of privacy in the process of digitalization in agriculture. This chapter was originally published at *The 21st Privacy Enhancing Technologies Symposium* [1] (Paper A).

Author Statement This paper constitutes a joint work of Sebastian Linsner, Franz Kuntke, Enno Steinbrink, Jonas Franken, and Christian Reuter. As corresponding and leading author, Sebastian led the overall research design, management, and writing process of the paper. The literature was collected by all authors, of which Jonas contributed a major part. The research design and choice of study design were done by Sebastian and Franz together. Sebastian and Franz planned and conducted the interviews and the clustering of results. The results and discussion were written by Sebastian and Enno, where Sebastian focused on the agricultural part and the findings regarding privacy, and Enno on the implications for research. The central implications of this work were mainly derived by Sebastian. Christian was a general advisor of this work and contributed with continuous feedback during all phases of the paper writing process.

Chapter 7 (*Paper B: Blockchain in Agriculture 4.0 – Empirical Study on Farmers' Expectations Towards Decentralized Services Based on Distributed Ledger Technology*) Digitalization, which proceeds in all branches as well as in agriculture through the use of new technology, sensors, and networking, requires the responsible handling of data. One possibility to manage data and extract value from it is to use blockchain technology. It is primarily enforced by the food industries and consumers to ensure traceability and transparency. In order to ensure a meaningful use of blockchain technology for agriculture, this application domain must also be considered and analyzed in terms of its social and business structures. This paper presents the results of a qualitative study in which 41 stakeholders from the agricultural sector answered questions in a written interview and participated in focus groups. It was found that it is attractive for farmers to adapt to new markets and technologies early on in order to gain a competitive advantage. However, their concerns about losing long-time local business partners and the social fabric in which they grew up must be taken into account. This chapter was originally published at *Proceedings of Mensch and Computer 2019* [2] (Paper B).

Author Statement The paper is a joint work of Sebastian Linsner, Franz Kuntke, Gina Maria Schmidbauer-Wolf, and Christian Reuter. As corresponding and leading author, Sebastian designed the structure of the paper and the contents. Gina Maria advised Franz and Sebastian regarding the design of qualitative research. The interviews were conducted by Sebastian and Franz. The coding of results was done by Sebastian Linsner and afterwards checked by Franz. Christian supported especially in refining the literature review, discussion and conclusion.

Chapter 8 (*Paper C: Supporting Users in Data Disclosure Scenarios in Agriculture through Transparency*) Collaboration in the era of digital transformation requires the exchange of operational data. Since data are hardly controllable once they have been published or shared with others, it is highly important that users are clearly informed about who has access to which data and how certain settings can prevent the disclosure of sensitive data. However, giving end users more control over their data through increased transparency could also lead to information overload. This is particularly true in the field of agriculture, where tight schedules put pressure on employees of small enterprises. We conduct an empirical prestudy with 52 German farmers to investigate current data sharing scenarios. From these insights, we derive requirements and a concept for data sharing solutions providing data flow transparency for users. To investigate the behavior of users and the effects of transparent UI controls, we evaluate a prototype with 18 persons. Our evaluation shows that farmers demand flexible and secure tools that adjust to their workflows. Also, data should be stored and processed locally, granting farmers data sovereignty. The evaluated transparent controls for data disclosure are easy to use and raise the users' awareness. However, the controls come at the cost of additional effort. This chapter was originally published in the journal *Behaviour & Information Technology (BIT)* [3] (Paper C).

Author Statement This paper constitutes a joint work of Sebastian Linsner, Enno Steinbrink, Franz Kuntke, Jonas Franken and Christian Reuter. As corresponding and leading author, Sebastian led the overall research design, management and writing process of the paper. The literature was collected by all authors, of which Jonas contributed a major part. The research design and choice of the study design for the pre-study were done by Sebastian and Franz together. Sebastian and Franz planned and conducted the interviews and the clustering of results. Afterwards, the results were integrated into the pre-study section by all authors. The implementation of the prototype was done by research assistants under the supervision of Sebastian. Enno contributed to the design of the study design and the questionnaires for the evaluation of the prototype, which was conducted by Sebastian. Results and discussion were written by Sebastian and Enno, where Enno focused on the statistical analysis of the questionnaires, while Sebastian focused on addressing the challenges derived from the pre-study. The central implications of this work were mainly derived by Sebastian. Christian was a general advisor of this work and contributed with continuous feedback during all phases of the paper writing process.

The second segment is comprised of publications about data sharing in the private context.

Chapter 9 (*Paper D: Support Personas: A Concept for Tailored Support of Users of Privacy-Enhancing Technologies*) In many applications and websites people use in their everyday life, their privacy and data is threatened, e.g., by script tracking during browsing. Although researchers and companies have developed privacy-preserving technologies, they are often difficult to use for lay users. In this paper, we first conducted a literature research to classify users into different support personas based on their privacy knowledge and privacy concern. The identified support personas allow developers to adjust their software to the support needs of different users. Second, taking into account the support personas, we developed PrivacyAssist, a browser extension which supports users with the issue of script tracking. In workshop sessions with 15 participants, we elaborated the concept of the browser extension on the basis of the results of our literature work, existing applications, and user requirements. We then developed the browser extension which provides different user interfaces for different support personas based on our participatory design approach. The qualitative evaluation with 20 participants attested PrivacyAssist a good user experience and revealed user preferences in terms of the displayed information. Thus, we conclude the concept of support personas is useful to develop usable applications that also educate the user. This chapter was originally published at *The 24th Privacy Enhancing Technologies Symposium* [4] (Paper D).

Author Statement This paper constitutes a joint work of Kilian Demuth, Sebastian Linsner, Tom Biselli, Marc-Andre Kaufhold and Christian Reuterwithin. As corresponding and leading author, Kilian led the overall research design, management and writing process of the paper. The authors contributed the literature review together where Kilian took the most work. The research design and choice of the theoretical model was done by Kilian and Sebastian, who were assisted by Tom.

Tom contributed to the statistical analysis and the generation of the study items. Kilian, Tom and Sebastian planned the interviews, which were conducted by Kilian. The results and discussion were mainly written by Kilian. The central implications of this work were mainly derived by Kilian. Marc-Andre and Christian were general advisors of this work and contributed with continuous feedback during all phases of the paper writing process.

Chapter 10 (*Paper E: Building Trust in Remote Attestation through Transparency – A Qualitative User Study on Observable Attestation*) In recent years, *Internet of Things (IoT)* devices have become increasingly important within the smart home domain. Given their widespread use, the security of the devices is a critical aspect. In particular, the vast majority of *IoT* devices are black-box systems running closed and pre-installed firmware. This raises concerns on the trustworthiness of these devices, especially considering that some of them are shipped with a microphone or a camera. Remote attestation aims at validating the trustworthiness of these devices by verifying the integrity of the software and sometimes hardware configuration. However, remote attestation is not transparent to the user since the user cannot validate whether the attestation has actually taken place and has not been manipulated by an attacker. In this paper, we conducted a qualitative user study with 35 participants to find out whether users trust the attestation process and whether the trust in the process can be improved by additional explanations in the application. We developed an application that allows users to attest their smart speaker using their smartphone over an audio channel to identify the attested device and observe the attestation process. In order to observe the differences between the application with and without explanations, we performed A/B testing. We discovered that trust increases when additional explanations of the technical process are provided and thus the understanding of the attestation process is improved. The enhanced understanding of the process also decreases the false trust many users have in devices or processes which they do not understand. Observable attestation via the audio channel also strengthens trust in the attestation process. We also found that the process of the observable attestation increases the trust in the device and the sense of security positively. Thus, increased transparency in security mechanisms increases not only the understanding but also the trust of users and reduces false assumptions about the own security. This chapter was originally published in the journal *Behaviour & Information Technology (BIT)* [5] (Paper E).

Author Statement This paper constitutes a joint work of Sebastian Linsner, Kilian Demuth, Sebastian Surminski, Lucas Davi and Christian Reuter,within.As corresponding and leading author, Sebastian Linsner led the overall research design, management and writing process of the paper. The authors contributed the literature review together where Kilian and Sebastian Surminski took the most work. The research design and choice of the theoretical model was done by Sebastian Linsner, who was assisted by Kilian. Sebastian Surminski developed and manufactured the demonstrators. Kilian made the adjustments of the application for the evaluation. Sebastian Linsner planned and conducted the interviews, while Kilian made the observations. Kilian was mainly responsible for data processing of the quantitative data and statistical analysis. The results and discussion were mainly written by Sebastian Linsner and Kilian. The central implications of this work were mainly derived by Sebastian Linsner. Lucas and Christian were general advisors of this work and contributed with continuous feedback during all phases of the paper writing process.

Chapter 11 (*Paper F: Decision-based Data Distribution (D^3): Enabling Users to Minimize Data Propagation in Privacy-sensitive Scenarios*) In many scenarios users have to communicate sensitive data with third parties such as doctors, lawyers, insurance companies, social workers or online shops. Handing over personal data is necessary to use those services but the delegation of tasks to increase efficiency still poses the risk that personal data might be leaked. To minimize this risk and further enhance the privacy of users, we propose an interaction concept that uses layered encryption of messages to provide a trade-off between privacy and usability. Users can choose which data is additionally encrypted in an inner layer, say, for the eyes of their doctor only, and which data is available in an outer (encrypted or unencrypted) layer for all staff members. Another benefit is the hiding of sensitive data from package inspection or crawling algorithms over emails, while less critical parts can still be processed by these systems via the partial access. To investigate this concept, we derive relevant use cases for form-based communication over email from a quantitative pre-study with 1011 participants, showing that general practitioners are the most suitable use case. We developed demonstrators for this use case and evaluated them in a qualitative study with 42 participants. Our results show that the possibility of minimizing the propagation of sensitive data through additional encryption is highly appreciated and the usage of form-based communication is a promising approach for the digital transformation. This chapter was originally published at *The 24th Privacy Enhancing Technologies Symposium* [6] (Paper F).

Author Statement This paper constitutes a joint work of Sebastian Linsner, Kilian Demuth, Marc Fischlin and Christian Reuter, within. As corresponding and leading author, Sebastian led the overall research design, management and writing process of the paper. The authors contributed the literature review together. The introduction was mainly constructed by Marc, who also provided the technical consulting regarding protocols. The research design and choice of the theoretical model was done by Sebastian with assistance of Kilian. Sebastian planned and conducted the interviews, while Kilian made the observations. Kilian was mainly responsible for data processing of the quantitative data and statistical analysis. The results and discussion were mainly written by Sebastian with Kilian assisting him. The central implications of this work were mainly derived by Sebastian. Christian was a general advisor of this work and contributed with continuous feedback during all phases of the paper writing process.

Research Design 2

This chapter presents important concepts and summarizes the methods that were applied within the papers found in the second part of this thesis.

2.1 Basic Concepts

This section presents the basic concepts that are used through-out this thesis to provide sufficient context, because some of these terms are used differently depending on the discipline.

Definition of Security Whenever the term "security" is used in this thesis, it refers to information security [26]. Confidentiality, integrity and availability of data have to be ensured and the hardware that is responsible for usage, storage and transmission of data. In the context of this thesis the concept of security is used to describe measures that prevent malicious parties from getting access to data or learning private information from observing the usage or the network.

Definition of Privacy Barkhuus [27] describes privacy as the preference of not sharing data with someone to either keep the data secret ("concern about others knowing something") or the wish to prevent burdening other with "seemingly insignificant information". Furthermore, personal privacy is the describes as the right to hide, "placing oneself in seclusion, away from self-selected people". In the context of this thesis this right to decide over the dissemination of personal data is extended to operational data. The reason behind this the context of SMEs that are

S. Linsner, *Privacy Preserving Data Management*, Technology, Peace and Security I Technologie, Frieden und Sicherheit,
https://doi.org/10.1007/978-3-658-47445-4_2

investigated. The owners of these businesses are the initial owners of the data and are often forced to share it with business partners. Furthermore, the operational data makes the owners of SMEs identifiable. Especially in agriculture, operational data refers to pieces of land, buildings or products. For big companies this is not an issue of privacy, but for SMEs, especially the family-run businesses each piece of operational data makes them identifiable as a person. Therefore, the need for privacy and data sovereignty is extended to operational data as well in the context of this thesis.

Definition of Data Sovereignty The concept of data sovereignty is heavily dependent on the context. Hummel, Braun, Tretter, and Dabrock [28] conducted a meta study on this topic and present different definitions from several domains. For this thesis the interpretation as an ability is most relevant: For this purpose the definition of Alboaie and Cosovan [19] is used: data sovereignty is "the ability of the user to have full control over his data".

2.2 Methods

After clarifying the context of this thesis, the methods used in the underlying publications are summarized. In Table 2.1, an overview of the methods for each paper is provided. Furthermore, for all these methods, it was crucial to get permission from the university's ethics board as participants' personal data was gathered and processed.

Table 2.1 Methods in the papers of this dissertation

Methods	Paper A	Paper B	Paper C	Paper D	Paper E	Paper F
Quantitative Studies						✓
→ *Survey*						✓
Qualitative Studies	✓	✓	✓	✓	✓	✓
→ *Interviews*			✓	✓	✓	✓
→ *Focus Groups*	✓	✓		✓		
→ *Usability Evaluation*			✓	✓	✓	✓

2.2.1 Quantitative Studies

Quantitative studies such as surveys are useful to validate hypotheses with a representative sample. Due to the high number of participants, the results are deemed very reliable, and statistical analysis can be performed. However, this form of research has some limitations: The hosting of a survey via a panel provider requires the use of validated items. Therefore, quantitative studies are not suitable for idea generation and the exploration of a research topic. Furthermore, due to the remote nature of these studies and the absence of a researcher conducting the study, it is hard to assess the validity of the data. The participants might just randomly answer questionnaires or provide false information. This can be minimized by attention checks and inverted items, but not mitigated. Overall, quantitative studies are good for confirming hypotheses with a representative sample, but they are not suited for hypothesis generation.

Paper F utilizes a quantitative survey (n = 1011) to find the most suitable use case for the main evaluation and to provide a basic set of data to which the qualitative sample can be compared. This method can be used to estimate the validity of the smaller qualitative sample and possible implications of the main results.

2.2.2 Qualitative Studies

Qualitative studies are usually of a smaller scale than quantitative studies. This implies that the results might not be projectable to a larger population or to other cultures. However, the different types of qualitative research provide much richer feedback on the study subject. Another advantage is that qualitative methods can be applied very early in an iterative design process, allowing for quality assurance and hypothesis generation. Furthermore, in most cases, researchers are present to conduct the study, allowing them to estimate the validity of the answers. On the other hand, the presence of the researcher can introduce biases by nudging the participants towards desired results or suggestive study material. Therefore, it is highly important to test and verify the study material before interacting with real participants. In the following, three different approaches to qualitative research are presented.

2.2.2.1 Interviews

In the context of this thesis, interviews [29] were utilized to accompany other qualitative methods. Papers C, D, E, and F used interviews as a final part of an evaluation to reflect on experiments, discuss guiding questions, gain feedback, or verify results from other parts of the current study. Interviews were mostly conducted in person, except the interviews in Paper C: These were conducted remotely due to the Covid-19 pandemic.

2.2.2.2 Focus Groups and Workshops

Focus groups [30], [31] are a special form of interview where multiple participants are interviewed simultaneously. This allows them to interact with each other and discuss relevant topics. The advantages of this method are the ability to interview many participants in a short amount of time and the interactions among the participants. Possible downsides originate from group dynamics, such as participants who are intimidated by others and do not participate as actively as in a single-person interview.

This method was applied in Paper A and Paper B to conduct expert interviews. The digital transformation of the domain and the introduction of new technologies such as blockchain was investigated and the discussion of the participants was beneficial to get different points of view on these topics. Furthermore, the discussion among the participants brought up issues and implications that would be neglected in an interview.

During the conduction of focus group interviews, we observed that some participants were more active than others. This might be due to personality or group dynamics. For Paper B, we added an additional task: Each participant was asked to write a short essay or statement on the topic of the study. This way, we allowed everyone to state their opinion without group pressure or other biases. This gave every participant a voice, even those who were reluctant to speak in the focus groups.

Another form of qualitative study that involves multiple participants is the conduction of workshops. In the context of this thesis, this method was applied in Paper D. In the iterative process of developing personas, workshops were held as a form of participatory design [32]. During the workshops, the mental models of users were investigated, and they were able to define solutions based on their needs and preferences. The results from the workshops were then used to derive personas, templates of user groups that define competencies, and needs of the corresponding user group.

2.2.2.3 Usability Evaluation

Usability evaluations are useful to investigate the effectiveness and usability of demonstrators such as prototypes and mock-ups. The researchers can witness the interaction of the participant with the demonstrators and investigate factors such as gestures, facial expressions, hesitation, and the time users need to complete a task. In doing so, they can find problems that would be unidentified by questionnaires and other quantitative methods. It is important that the researchers do not interfere with the experiment to avoid biases. Therefore, the presence of multiple researchers during the evaluation can minimize personal biases and assure the quality of the study.

Most usability evaluations in this thesis were comprised of hands-on experiments accompanied by questionnaires. The questionnaires served multiple purposes: On the one hand, they were used to measure the usability of the demonstrator with standardized scales such as UEQ-S. On the other hand, further data could be gathered to describe the sample better, such as demographic data and scales measuring privacy competence or privacy concerns. By providing more information on the sample, the limitation of non-representativeness due to the smaller sample size can be evaluated better.

Findings

3

This chapter summarizes the findings of the publications which can be found in Part II of this thesis. Each publication is analyzes regarding its contribution to each of the three research directions which resemble the three sub-questions of this thesis (see Fig. 3.1).

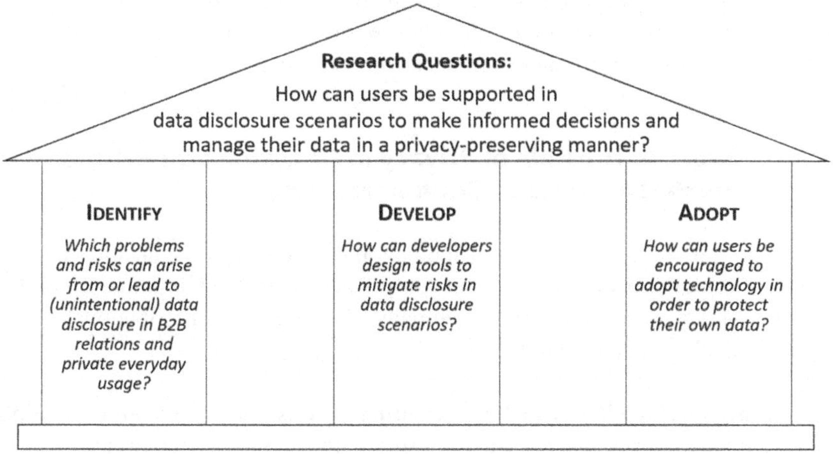

Fig. 3.1 Representation of the research questions of this thesis

S. Linsner, *Privacy Preserving Data Management*, Technology, Peace and Security I Technologie, Frieden und Sicherheit,
https://doi.org/10.1007/978-3-658-47445-4_3

In the paragraphs on IDENTIFY, all findings are summarized which point out risks and problems that result from data disclosure or could lead to it. Likewise, the paragraphs on DEVELOP present all measures and recommendations for developers that help them to create tools that assist users in managing their data in a privacy-preserving way. Finally, each publication is analyzed for techniques to bring security- and privacy-preserving technologies into practice. these findings are presented in the paragraphs on ADOPT.

3.1 Segment I: Assisting Users in B2B Data Disclosure Scenarios

In this segment, the focus is on the support of users who have to disclose data in order to carry out business operations. This is especially important in domains where collaborative work and division of work are inevitable. The three publications contributing to this segment provide different views on the situation in agriculture as a representative domain currently undergoing the process of digitalization. Besides broad overviews of the situation in the whole domain, some technologies and trends are highlighted and investigated with a narrower perspective.

3.1.1 Paper A: The Role of Privacy in Digitalization—Analysing the Perspectives of German Farmers

This paper provides an overview of the ongoing process of digitalization in agriculture and highlights the implications on privacy, as well as its influence on the adoption of new technologies in this domain. The amount of contributions to each research focus are depicted in Fig. 3.2.

Contributions to the Research Focus Identify In this study we identified multiple issues which are connected to data disclosure in the domain of agriculture.

Technical aspects: By introducing digitalization to agriculture, new points of failure and dependencies arise. Workflows have to be modified and can be disrupted by technical faults and errors. Furthermore, the work profile within SMEs is changing, when the business owners have to perform more and more digital tasks, which may differ significantly from the actual profession they work in. Often the promised increase in work efficiency due to digital tools is nullified by the extra office work. Another technical issue arising from technical aspects is the possible vendor lock-

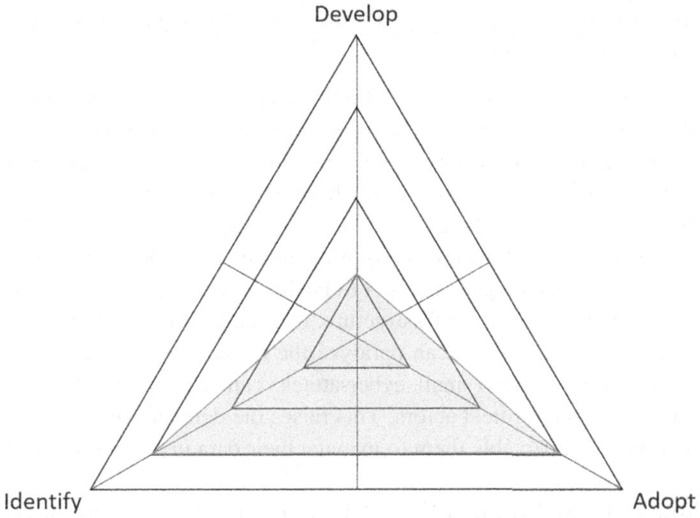

Fig. 3.2 Contributions of Paper A for each research focus

in effect due to lack of compatibility. Forcing users to stick to one company's solutions is not only expensive for SMEs who have to buy specific products but can also force users to accept terms of data sharing due to the lack of alternatives. This poses further risks: Companies can easily profile their customers, granting them huge advantages over them. A major fear of farmers is that they will face personalized prices if companies get too much of their data: If suppliers and buyers can estimate the farmers' production cost and the estimated yield they can adjust their prices to reduce the margin of the farmers. Furthermore, the perception of data as a resource plays a role: Knowing about the ability to sell data to third parties and the advantages of data collection for companies, farmers begin to realize that data is not a byproduct of digitalization but a resource itself. Therefore, they want business concepts to benefit from the value of their own data if they have to share it.

Social aspects: One surprising fact that arose during the interviews was that digitalization and especially security leave an unpleasant aftertaste for farmers regarding the perception of their business partners. In agriculture, business contacts are long grown relationships, lasting for generations. By introducing the concepts of security and privacy into this domain, the general perception of trust toward each other is disrupted. Furthermore, by shifting the interaction with business partners from the real world to online interactions, social contacts suffer. To conclude, designing data disclosure restrictive to enhance the sovereignty of the data owners can provoke mistrust and social stigmata.

Ethical aspects: This category includes findings that cover aspects of data collection that are legal but undesirable from the perspective of users, especially SMEs. One major fear is that data disclosure grants companies a massive advantage over the framers, eventually leading to the point where the farmers are bought out, and the fields are worked by using the collected data of the previous owners. This would force many SMEs out of business, disrupting entire areas with severe social problems. Furthermore, reducing the number of businesses in this domain is detrimental to the resilience of this critical infrastructure. Digitalization poses the threat of malware and cyber-attacks, which can paralyze the entire business. If the number of businesses in a domain is too small, cyber-attacks can cause serious harm to the entire sector by targeting critical actors. This raises the demand for solutions to keep SMEs operational and enable them to manage their data in a sovereign manner.

Contributions to the Research Focus Develop Since larger companies upstream and downstream of the supply chain are the main driving force behind technological innovation in this domain, farmers are underrepresented in the development process. Some technologies claim to provide benefits to the farmers, such as opportunities to trade their goods online over virtual marketplaces or Blockchain environments. However, what sounds like more control over their data and products can lead to problematic market situations because production standards (and costs) differ globally, putting farmers at a disadvantage. The truth is that the ones who benefit from this technology are not the farmers but retailers who want to offer increased traceability to their customers. To conclude, to improve the situation of SMEs in data disclosure scenarios, new technological solutions should consider the impact and possible downsides of their adoption for the users and the impact on the domain.

Contributions to the Research Focus Adopt The first major factor for the adoption of new technologies in agriculture is the dependency on the benefits that are provided by this technology. The adoption of new machines or digital tools is expensive, and the expected benefits have to compensate for this. On the other hand, possible disadvantages, such as privacy concerns, reduce this perceived benefit and inhibit the adoption of new technologies. This is reflected in different sub-domains of agriculture. Farmers who breed cattle for different purposes not only benefit from advancements in digitalized equipment, but they are also dependent on mechanized farm equipment to guarantee the well-being of their cattle (e.g. milking robots) or adhere to laws and regulations (e.g. ventilation systems, generators to compensate power outages, etc.) issued by governmental authorities. Grain farmers are not necessarily reliant on digital tools but can benefit from a huge optimization potential through precision agriculture. Viticulture, on the other hand, benefits least from digitalization since many work steps are still carried out manually. Administration and retail can be optimized via digitalization, which makes wine farmers relatively free to choose whether they want to adopt new technologies or not. Their cost-benefit calculation is, therefore, more heavily influenced by privacy concerns and security evaluations. This kind of calculation is not limited to agriculture. Most enterprises use risk calculations when investing in security mechanisms. By evaluating risks by their severity, likelihood, and cost to compensate for an incident, the decisions to invest in security mechanisms are made.

Another factor for adopting new technology is more specific to agriculture or domains which receive public funding and subventions. These funds are often tied to regulations which businesses have to adhere to. By digitalizing the work processes, the retraceability is increased, and violations can be found more easily. However, this leads to a feeling of governmental surveillance and the fear that technical errors like false sensor readings can lead to sanctions. Therefore, becoming transparent for others is an inhibiting factor to technology adoption.

In conclusion, the results from this paper make it clear that privacy perception is influenced by dependencies arising from the use of new technology, which, in turn, are important factors in the decision of whether to adopt a new technology or not. Thus, technology adoption shapes privacy perception, but privacy concerns affect technology adoption.

3.1.2 Paper B: Blockchain in Agriculture 4.0

After presenting the situation for owners of SMEs in agriculture, the process of digitalization in this domain and general issues regarding the handling of data, the following publication focuses on one particular technology, its introduction to the domain, and the resulting implications. Despite having a narrow focus, the results also apply to other technologies and domains. The amount of contributions to each research focus are depicted in Fig. 3.3.

Contributions to the Research Focus Identify Before digitalization, malicious actors had to break into buildings or trespass on a company's work area. However, when adopting digital tools, this attack surface increases. Furthermore, due to the exposed position on fields, sensors and other equipment that is connected to a Blockchain is prone to physical access by attackers. This is one of the strongest

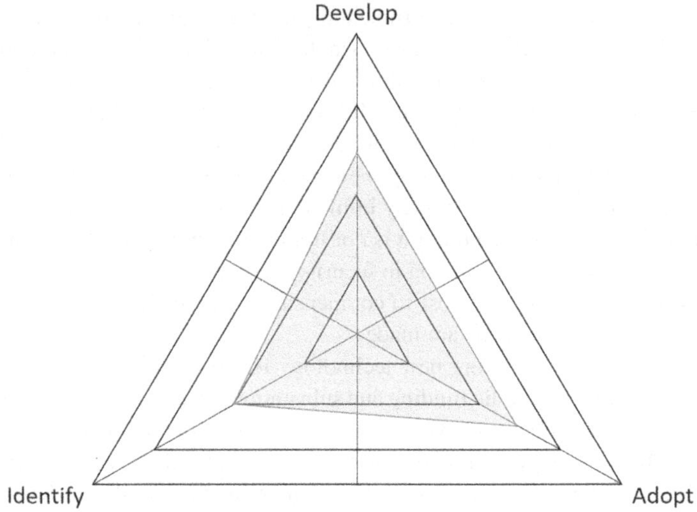

Fig. 3.3 Contributions of Paper B for each research focus

assumptions in risk assessment and requires extensive security measures. Security measures have to be taken, and knowledge of potential risks has to be acquired.

Even in a secure setting, trust is needed: Many farmers lack trust in sensors and Distributed Ledger Technology (DLT). By making simple errors, they might get into trouble because they cannot adhere to regulations or quality criteria. Furthermore, governmental surveillance was a concern. However, state actors are not the only problem for farmers: By disclosing data for reporting duties, unintended data leakage may occur. Farmers have to disclose certain data to prove adherence to quality criteria in retail or regulations towards authorities. However, they do not want this data to end up in the hands of their suppliers and buyers. Those actors could otherwise predict the farmers' yield and offer them personalized prices, which leads to price dumping. Therefore, binding regulations have to be made to avoid data misuse and data leakage.

Contributions to the Research Focus Develop The development of Blockchain technology, or DLT in general, led to new business models in the agricultural domain. These claim to increase security and food safety but can be a double-edged sword. Those new business models include direct retail for farmers to circumvent local traders, who may have a market monopoly and dictate prices. However, this could lead to competition not only among regional farmers but also among global players. Since production standards and costs differ between countries, this imbalance could lead to price dumping and unfair competition. Another business model is the trading of data itself. Data is becoming increasingly important, and its collection and processing have huge benefits for many companies. Therefore, paying the data owners for their contribution to datasets could be a new business model. However, many farmers are skeptical about whether the compensation they receive is worth the benefits they grant other players alongside the supply chain.

Another technical perspective is provided by privacy-enhancing technologies for DLT. By providing transparent documentation of the whole production process and all actors, food items can be easily traced, and quality standards can be verified. However, permanent documentation not only harbors risks for the data owners but also violates privacy laws, which require pre-defined deletion periods. Technical solutions are storing encrypted data on the Blockchain and deleting the decryption key after the deletion period expires. Another approach is the off-chain storage of data while only hash values are stored on the chain. The benefits of this approach are smaller data size on the chain and the possibility of deleting the original data if necessary.

In summary, there are different technologies that try to increase security, safety, and privacy in the DLT setting. However, these are solutions for problems that only arise through the adoption of the technology itself.

Contributions to the Research Focus Adopt Important factors for the adoption of technology and new data-driven business models are diverse. First, there is a social or traditional factor. Business collaborations and partnerships are well-established and sometimes last for generations. This does not only provide security and plannable work steps within collaborations but is often considered as a social interaction with well-trusted partners. On the other hand, new globalized business models with direct retail and data selling have not earned this trust yet and may bring risks for businesses.

The second factor is the direct and indirect cost of adoption: Besides the initial cost of acquiring new technology and the continuous cost of running the needed equipment, there are additional hidden costs. First of all, there is the need to educate the operators. In agricultural SMEs, this is a whole new field of work which has to be learned and mastered to a degree where it is safe (and secure) to operate. Furthermore, there is the investment in the form of time. Managing all data, preparing datasets for the machines, and adhering to reporting duties cause a lot of office work, which is a burden for SMEs, which often operate at their maximum capacities.

Therefore, the initial and the resulting cost are a serious hindrance especially for SMEs to adopt new technologies.

3.1.3 Paper C: Supporting Users in Data Disclosure Scenarios in Agriculture Through Transparency

The previous two publications pointed out that lacking privacy is one of the most relevant reasons for the hesitant adoption of new digitized tools in agriculture. To combat this issue and to grant farmers control over their data and its disclosure to others, a concept for data management within a supply chain is presented by the following paper. The goal was to provide farmers with a tool that enables them to minimize data propagation in collaborative scenarios. In agriculture, multiple actors are required for specific work processes. For example: A seeding operation requires the suppliers to deliver seeds to the correct field. The machinery ring has to find suitable contractors. The contractors need data on the fields, the historic operations in the field, and the current operation. The farmer has to supervise and manage all actors, supply them with the needed data, and coordinate previous work steps on the field. To give the farmers more control of their data, we developed a tool that contained a form for specific work orders, offering data fields for all needed data and control mechanisms to specify who could access this data. Based on this control mechanism, the views of the other actors were generated, assuring that only the disclosed data leaves the computer of the farmer and is transmitted to the corresponding recipient. This setup was evaluated with actors from agriculture and a non-agricultural control sample. The amount of contributions to each research focus are depicted in Fig. 3.4.

Contributions to the Research Focus Identify In the preparation and during the evaluation, several problems regarding data disclosure scenarios were identified: In the preparation for the development process of our tool, we encountered the so-called control paradox in literature. It indicates that users tend to disclose more data when they have more control over the disclosure process. This was alarming for us since our aim was to make the data distribution more transparent to the farmers and enable them to customize the data disclosure to their needs. During the study, the trade-off between privacy and usability was discussed. Some participants stated that they were afraid of being too restrictive with the data disclosure because this could result in a failed job execution due to lacking data. In real-world scenarios this problem could be solved with requirement fields. Furthermore, it seems to be an issue that is not only related to agriculture (see Paper F).

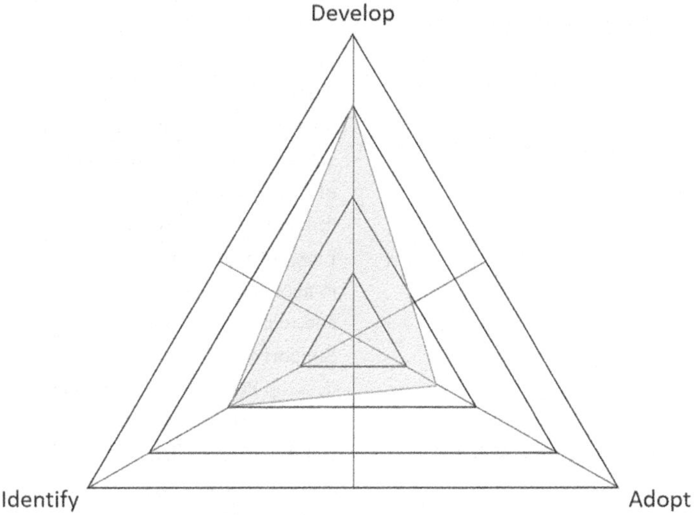

Fig. 3.4 Contributions of Paper C for each research focus

Contributions to the Research Focus Develop In order to grant users more control over their data, we aimed to make the data distribution more transparent. This was a challenge because literature indicates the existence of a privacy paradox which counteracts our approach. Another factor to consider was information overload. Making the process transparent can enhance the understanding, but could lead to an overload of supplied information, negating the educational effect.

From our own pre-study, we derived 5 challenges for the implementation of digital tools in agriculture, which are shown in Table 3.1.

Contributions to the Research Focus Adopt A major factor in the adoption of such user-empowering tools is time. The participants in our study reported time constraints in their work and that detailed planning for data distribution would be a burden. They demanded countermeasures such as global settings for the recipients or the possibility to pre-fill the form with data from last year's job order. They were willing to do the detailed planning once if they could reuse their setting for each recipient in future work orders.

Table 3.1 Challenges derived from Paper C

Challenges	Platform Requirement
Challenge 1: Compatibility	Application of standard data formats to facilitate compatibility across products. Independence of internet connectivity through local accessibilty.
Challenge 2: Sector Diversity	Flexibility of the platform to account for different actors and roles.
Challenge 3: Media Diversity	Support of interfaces on desktop PCs and mobile phones as most commonly used media.
Challenge 4: Transparency Balance	Improvement of data sharing choices and provision of job specific approvals. Restrictive pre-selection to meet privacy by default.
Challenge 5: Retraceability	Ex-post traceability of data sharing decisions and contents provides evidence of all transactions for legal proceedings.

3.2 Segment II: Assisting Users in Private Data Disclosure Scenarios

The previous segment presented the findings regarding data disclosure scenarios in B2B relations, illustrated by the investigated domain of agriculture. This segment focuses on individual users as a target group to investigate their needs in data disclosure scenarios. To accomplish this, three approaches are presented: Paper D investigated the needs of different user groups and derives personas from them. Paper E utilizes transparency as measure to educate users and researches the connection of understanding and trust. Finally, Paper F explores two approaches of enabling users to define data flows in data disclosure scenarios.

3.2.1 Paper D: Support Personas: A Concept for Tailored Support of Users of Privacy-Enhancing Technologies

In this paper, support personas are explored as a user-centered intervention to assist users in privacy-sensitive contexts. By measuring the level of needed assistance, each user group can be provided with tailored support. In this case, different UIs were provided to assist users in the context of script tracking in browsers. The amount of contributions to each research focus are depicted in Fig. 3.5.

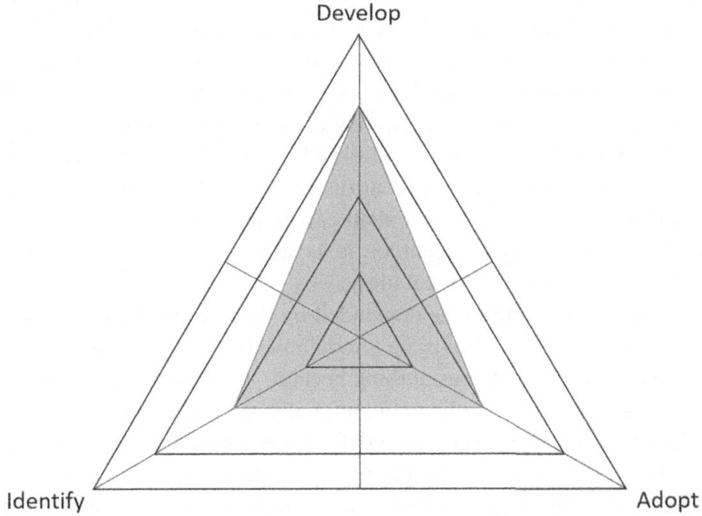

Fig. 3.5 Contributions of Paper D for each research focus

Contributions to the Research Focus Identify From the literature and the workshop sessions, five challenges were deducted (see Table 2 in Paper D): (1) Give the user confidence in own capabilities; (2) Prevention of user frustration; (3) Easy to use and understandable UI; (4) Provide more detailed information to experts and (5) Freedom of choice. To reduce redundancy, the explanation of each challenge is moved to the following subsection where the recommendations for developers are presented to meet the needs of the users.

Contributions to the Research Focus Develop For the creation of support personas the approach of participatory design was chosen. In computer science, an iterative development process is recommended to fit the product to the target group. However, the participatory design approach involves users in the earliest stages of development to use their feedback in the design stage. In the case of this publication, users were invited to workshop sessions to discuss different affordances and requirements for personas that reflect the need for support when dealing with data disclosure scenarios, in this particular case: browser tracking.

To address the first identified challenge, different UIs should be developed based on the privacy competence of the users. Providing each user group with tailor-made UIs prevents information overload and enables them to act according to their preferences and needs. Users with low privacy competence get basic functionalities to customize their experience, while expert users are provided with more detailed controls to fine-tune the settings. This differentiation is important because expert users might be uncomfortable with less control, while lay users might get discouraged when confronted with fine-grained controls for things they might not even know. This would lower their confidence in their capabilities and their perception of self-efficacy [33]. Additionally, information texts should be provided that inform the users with suitable extend and vocabulary.

To prevent user frustration, as demanded by the second challenge, secure and privacy-preserving settings should be the default. If users trust the tool to protect them, but in reality, the tool does nothing because it was not set up correctly, users will get disappointed and might unintentionally disclose data they thought was protected by the tool. However, those defaults should not limit the possibility of customization for the users, or otherwise these restriction can pose a burden for users, resulting in frustration and discarding of the tool. To further avoid "surprises" for the user, developers should research the mental models of their target group and invest in user education (e.g., through transparency).

This leads to the recommendations to address the third challenge: Transparency is a suitable intervention to make the tool more accessible and foster understanding of the process behind the tool. Increased understanding does not only build trust, but also mitigates usage errors. Educational measures such as transparency make tools easier to learn and therefore, easier to use. However, in conjunction with the fourth challenge, the amount of information should suit the target group or persona.

As already mentioned in the first challenge, it is crucial for the fourth challenge to provide experts with fine-grained controls to customize their tools. Experts have a higher privacy competence and are more informed about the risks that arise from online interactions. Furthermore, the higher competence allows them to evaluate these risks and the corresponding trade-offs better. This allows for informed decisions on how to customize the settings. Preventing them from customizing their tools and giving them only some pre-defined settings would discourage them from using the tool and lead them to search for alternatives.

Finally, regarding the fifth challenge, users should be able to choose the UI they prefer. Assigning them to a recommended UI could offend them and lead to reactance [23]. Such a paternalistic measure could be considered offensive and diminishes the feeling of self-efficacy [33]. Recommending a UI for them but allowing them to try out others and then choose their preferred version gives users a feeling of control which leads to a positive attitude towards the tool. Furthermore, users are individuals, and in the end, each user has individual needs. A persona is useful for considering different needs during development, but it is unlikely that there is a persona that reflects all users and thus covers all their needs. Therefore, giving them the freedom to choose their UI compensates for this inaccuracy.

Contributions to the Research Focus Adopt The previously stated recommendations to meet the challenges can help to reduce barriers to adoption. Especially the participatory design and the personas shall ensure, that the needs of the potential users are met. However, the customizability should not be neglected. Even if each user gets an optimal UI, the preferences could change over time. In the evaluation of the browser extension, it was found that some participants with high privacy competence preferred the detailed version of the UI, but also stated that they would prefer the extended view for everyday usage. This shows that preferences are also influenced by the context and can change due to external factors. This underlines the demand for highly customizable solutions that can adapt to the situation.

3.2.2 Paper E: Building Trust in Remote Attestation through Transparency—A Qualitative User Study on Observable Attestation

After investigating different user needs and deriving personas to assist developers in designing tools that enable users to interact with them in a way that suits their needs, this paper investigates different approaches to provide users with transparency. Transparency is a promising approach to increase the users' understanding and educate them. The following papers investigate how this raised understanding leads to increased trust. As use case, a novel approach of attestation was investigated alongside different methods to apply transparency to security features in order to increase trust in the system. The amount of contributions to each research focus are depicted in Fig. 3.6.

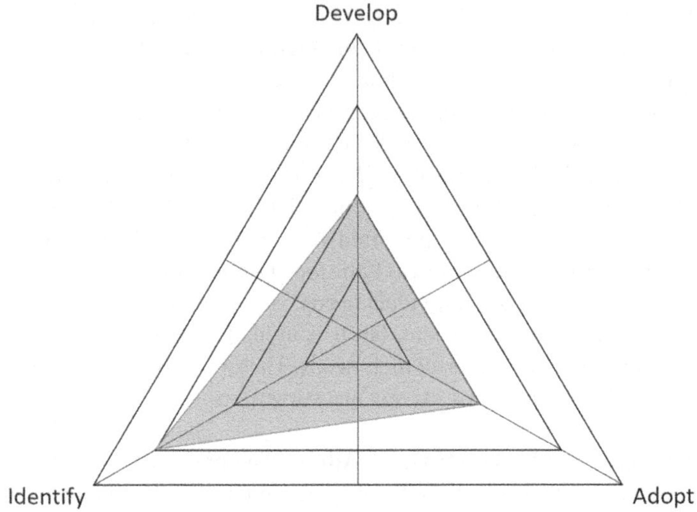

Fig. 3.6 Contributions of Paper E for each research focus

Contributions to the Research Focus Identify The main issues discovered by this study revolve around the mental models of users. For example, participants showed an understanding for the need of training data for AI training, but privacy incidents which became public raised their demand for higher security and privacy standards. Furthermore, companies should not use the data of their customers to train AI models without consent. Another issue with mental models that affect users' privacy behavior is resignation. Multiple participants stated that they believe that some big companies already have all their data, and therefore, they do not need to care about privacy, because it was already disclosed. Similar statements were made in the context of mobile devices such as smartphones. Participants believed that their phones were spying on their everyday lives, and thus, their data was already collected. These issues led to the perception that users have no power, resulting in resignation in the context of data sovereignty and privacy. In some cases, this led to a strange form of "trust": participants believed that their data was already collected by companies and since nothing "bad" had happened so far, they attested trustworthiness to the companies. The crude logic behind this is something like this: They have my data, and nothing bad happened; therefore, they probably treat my data correctly and deserve my trust.

Trust, in general, proved to be a very diverse topic in this study: Despite gaslighting themselves into "trust" towards companies, we discovered how (incorrect) mental models affected trust towards security features. In this particular study, the context was a new observable attestation scheme that allowed users to witness the verification process to increase trust in the system. However, this transparency could lead to disappointment. By educating the user about the attestation scheme, false mental models were revoked. This reduced "illusionary" trust that was misplaced in the security mechanism by false attribution by the users. For example, users initially thought that attestation would also prevent the manufacturer of a smart speaker from data collection. They were disappointed to learn that attestation "only" prevents data collection by malicious third parties. This showed that it is necessary to distinguish between different types of trust: real trust can be achieved by educating the users and providing them with transparency. By doing so, "illusionary" trust is reduced and false mental models revoked. This also is a desirable outcome, since false attribution towards security and privacy mechanisms can lead to a misleading feeling of security. Users might think that they are protected but might disclose data unintentionally due to overestimating the security and privacy features. To mitigate this risk, false mental models and the "illusionary" trust must be corrected for the correct use of privacy-enhancing technologies and security measures.

A third aspect revolving around mental models and trust in attestation was the handling of formerly corrupted devices. In the evaluation setup, a corrupted device was simulated. After resetting the device, the attestation should succeed. In the following interview, the participants were asked whether they would continue to use a device that was corrupted, reset, and successfully attested. Among the sample, two different attitudes could be observed: Some participants would lose all trust and immediately stop using the device, and probably all similar devices, too. Others said that as long as the attestation succeeded after reset, they would continue to use it normally but would attest the device more often to monitor the situation. Overall, these attitudes may originate from personal preferences and experience, but they also shed light on the level of trust toward attestation.

Contributions to the Research Focus Develop Transparency proved to be a suitable technique for increasing the understanding of users. From the two different methods used, the explanation text was more efficient than the info texts in each step. The latter also resulted in an increased understanding. However, due to the sample size, this difference was not statistically significant. The explanation text, which could be reached via the "explanation"-button on the main page, proved to be far more effective in educating the users. However, this method had a disadvantage compared to the transparency intervention in each step: the users who ignored the explanation would have missed out on transparency measures without the small explanations during the attestation process. For developers, it might be best to combine these approaches to reach more users. During the evaluation, some users who did not use the explanation text demanded an "?"- or "i"-button for additional information. This shows that the users' expectations could affect the effectiveness of transparency interventions: just because the explanation button was presented in a different way than the users were used to, they missed out on it. However, since this was a qualitative evaluation, other explanations might be possible due to the sample size. Alternatively, these users might have asked for an "?"- or "i"-button to justify that they could not explain the functionality of the attestation during the interviews and wanted to justify it by blaming the design of the demonstrator.

Contributions to the Research Focus Adopt Understanding the mental models of the target group is critical to bringing security and privacy-enhancing tools into practice. Users need a general understanding of why they should use these tools, which requires an awareness of potential risks. Furthermore, incorrect mental models could affect the adoption of these technologies: If users overestimate their protection, they are prone to unintentional data disclosure due to misuse of security and privacy mechanisms. Without the correct mental model, users might not even feel the need to adopt a technology. Therefore, the education of users is critical and should also be included in the advertisement strategy. For companies, it may be wise to invest in security and privacy measures such as attestation for their products because participants indicated that they would be willing to pay more for a product if it had built-in security features. The additional investment in security and privacy could be a unique selling point in a competitive market.

3.2.3 Paper F: Decision-based Data Distribution (D^3): Enabling Users to Minimize Data Propagation in Privacy-sensitive Scenarios

After providing users with usability through suitable UIs and education through different levels of transparency, controls are needed to enact their own decisions regarding data disclosure. The following paper presents an interaction concept that utilizes everyday use cases as opportunities to give users more control over the disclosure of their data. To further investigate, whether the increased control leads to less data leakage, two different demonstrators are developed: The first one gives the users the possibility to protect sensitive data with layered encryption by using a designated input field. The second demonstrator allows users to decide which input fields should be additionally encrypted, granting them full customizability. The paper investigates the possibility of using layered encryption in everyday scenarios to enhance the privacy and security of end users. By giving users an easy and understandable tool to use additional layers of encryption for the most sensitive parts of messages, they can act according to their privacy preferences and build trust towards the recipient. This feature was investigated in the context of medical offices, allowing users to decide which information is for the eyes of the doctor only and which information can be shared with other staff members. The amount of contributions to each research focus are depicted in Fig. 3.7.

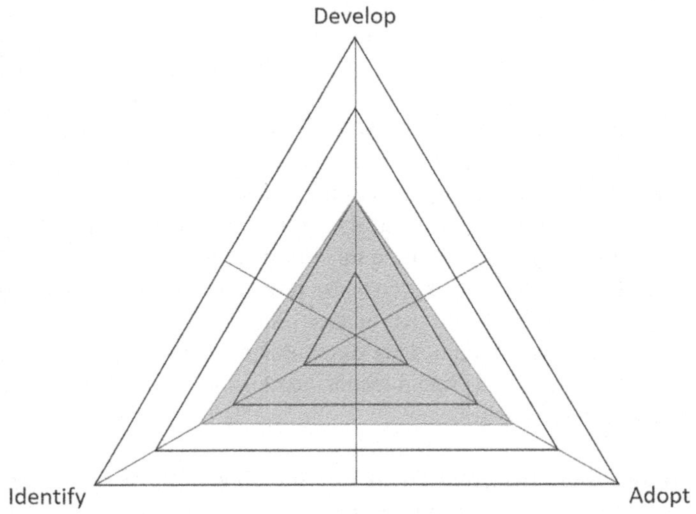

Fig. 3.7 Contributions of Paper F for each research focus

Contributions to the Research Focus Identify This paper was motivated by the problem that companies are suspected of sharing their secret keys with the middlebox or firewall for scanning incoming messages. This is a major issue for privacy-sensitive contexts and for trust in security features.

During the evaluation of the demonstrators, social issues were discovered. These could be potential barriers that prevent people from receiving medical aid. Among these were reasons like relatives or acquaintances working within the medical office and patients being embarrassed to disclose their health conditions to them. The participants stated that in such a case, they would seek out a different medical office, if available, or reported instances where relatives did not seek medical help at all in fear of rumors. Especially in rural areas with lower coverage of medical offices, this hinders people from getting medical help. Privacy-preserving online communication with doctors would be a help for these people.

Further issues were that bystanders could eavesdrop on patients stating their issues in the medical office or merely the fact that rumors might spread just because someone was seen entering a medical office. This could be avoided with extended online services, which is not exclusive to the presented interaction concept, but illustrates that this concept can provide benefits for all citizens and not only those, who care about privacy.

One particular issue arose from the adoption of seemingly non-related technologies: Due to the broad availability of social media and smartphones, established customs in medical offices could become problematic. An instance was reported where the usual interaction at the front desk caused problems: A female patient was asked to state her name and date of birth at the front desk to identify herself as a patient, which is common practice in Germany. However, due to the broad availability of smartphones and social media, this becomes problematic: A male bystander eavesdropped on it and used this information to find and contact her on social media. Thus, such practices can cause problems such as stalking or identity theft. Freely announcing private data makes users identifiable, not only to those to whom they are intended to be. This shows that even the adoption of seemingly unrelated technology, such as social media, can lead to problems when combined with problematic prerequisites.

Contributions to the Research Focus Develop One of the main findings of the development process of privacy-preserving technologies is that UI design decisions affect not only the usability of a tool but also the mental models of the users regarding security and privacy. Participants stated that the clear separation of input fields emphasized the feeling of separation between protected and unprotected data in the non-customizable version of the demonstrator. On the other hand, in the customizable version, participants praised the ease of use and that it was apparent how to protect each input field due to the adjacent placement of the controls. Furthermore, this design affected the mental models of the users in another way: Because they were able to control which data was protected and which data was not additionally protected, they had the feeling that everything they disclosed was disclosed by their choice. In other forms and situations where users had to disclose data, they felt like something was taken away from them. However, by giving them the possibility to use additional protection they felt like they were giving the data away based on their decision. This is a completely different perception of the interaction and the disclosure process, which leads to increased trust on the receiving side. However, this is not always a good thing: These design decisions are helpful to enable users and provide trust in the system. If used to benefit the users, these interventions enhance security and privacy. On the other hand, if used to deceive users, these design choices could lure users into a false sense of security and misplaced trust. It is the same problem as with other dual-use technologies: the intention of the designer decides whether a tool becomes a weapon. Furthermore, there is another risk: the so-called privacy paradox [34, 35]. From the observations during the interviews, one reason for the disclosure of data, despite the possibility of withholding the data, is increased trust due to the design that creates trust. By increasing the trust, the users perceive the receiving side as the "good guys" and feel more eager to put data in their hands.

Contributions to the Research Focus Adopt Using technology to rethink established systems can create benefits for security and privacy. In this particular use case, great enthusiasm was found for secure form-based communication with doctors and even other recipients. Beyond the primary use of secure and privacy-preserving communication, the proposed interaction concept offers great benefits for users. The possibility to use online forms to communicate with doctors was perceived as a major improvement since few doctors offer online services, and most of the patients have to use the telephone to reach out to the medical office. Especially for people with telephone anxiety, this is a huge problem. However, people who work during the opening hours of the medical office find it hard to get an appointment. Another group of patients who would benefit from this form of communication are those who have social barriers to seeking help. Some patients do not contact the medical office, either due to stigmatization by society or the fear of rumors among their social peers. The benefits of online communication with doctors are not new, but this paper underlines that the implementation of the interaction concept that provides more security and privacy does not come with a trade-off but is an additional driving force for digital transformation in this domain.

Integrating controls for additional protection into digital tools can benefit society: Those who do not care about privacy or just have never thought about it get confronted with the possibility of enacting sovereignty over their data. This way, an educational process can be started that increases the level of awareness within society overall.

Discussion

4

This chapter reflects on the findings of the publications in Part II and uses them to answer the overarching research questions of this thesis. For this purpose, each research focus is reviewed separately an afterwards, an overarching model (see 4.4) is formulated to answer the main research question.

4.1 Research Focus Identify

The first research focus is IDENTIFY. Research contributing to this research area revolves around finding problems and risks that either arise from data leakage or lead to (unintentional) data disclosure. In this way, the first sub-question of this thesis will be answered:

RQ1 **Which problems and risks can arise from or lead to (unintentional) data disclosure in B2B relations and private everyday usage?**

The findings that relate to this question are divided into three categories: (1) Technical aspects that arise from the adoption of new technologies and practices. (2) Social aspects and mental models that can be a hindrance to privacy-preserving technologies or directly affect the behavior of users. Finally, (3) ethical aspects are discussed that imply problems for society arising from data disclosure.

S. Linsner, *Privacy Preserving Data Management*, Technology, Peace and Security I Technologie, Frieden und Sicherheit,
https://doi.org/10.1007/978-3-658-47445-4_4

4.1.1 Technical Aspects

From the adoption of new technologies, new problems can arise. For example, in the interviews and focus groups with farmers (Papers A and B), the topic of new dependencies and points of failure was brought up. Both resulted from the adoption of new digitized farm equipment. Furthermore, due to incompatibility issues of the equipment, vendor lock-in effects could occur. These problems then lead to data disclosure because farmers are forced to use the data management tools of the vendors and lose sovereignty over their data, as Papers A and B showed. The data itself becomes a product that has to be passed down the supply chain. Without the needed data, machinery will not run, and collaborative jobs may fail. However, those who have to disclose data are often those who profit least from data as a resource. Furthermore, processing data to provide it to others does increase the workload. The farmers in the focus group stated that the increased efficiency by adopting digitized equipment [36, 37] is almost nullified due to the overhead in office work. This issue makes the data disclosure even less attractive since the workload increases without direct benefits or compensation. On the other hand, the disclosed data can be used and even misused by others along the supply chain [20]. To conclude, the adoption of new digitized tools and technologies comes with great risks of losing data sovereignty and unintended data disclosure.

This problem is not limited to businesses such as SMEs in agriculture. Adoption of new technologies can also pose risks to private users, raising the demand for new solutions and for re-thinking old habits. An example was presented in Paper F, where the widespread adoption of smartphones and social media led to unintended data disclosure through old customs, such as stating personal data in a medical office for identification.

Another problem arising from digital tools and equipment, such as sensors, becomes most apparent in the context of DLT, such as Blockchain (Paper B). Due to the constant disclosure of measured data, sensors that are connected to a blockchain not only provide other actors with a huge amount of data but, due to their immutability, pose a threat to farmers who are dependent on subsidies. Farmers have to adhere to governmental regulations and environmental standards to receive subsidies from their country or the EU. Sensor errors put these at risk, especially considering the possibility of malicious acts such as sabotage. Sensors in agriculture are extraordinarily prone to attacks since they are not deployed within a secured lab environment but are openly accessible in the fields. This grants potential attackers the ability to gain physical access, the strongest assumption in IT security assessment. In the context of smart phones the issue of "public devices" is investigated in research [38], however, this problem is often overlooked in IoT development, because most devices are build for indoor scenarios, neglecting the use cases that require outdoor deployment.

4.1.2 Social Aspects and Mental Models

Social surroundings seem to have a big influence on private users as well as B2B relations. This affects the usage of digital tools or the tendency to avoid them [23]. Papers A and B showed that long-term business relations are seen as social peers and that this relationship is disrupted by digital tools. A raised awareness for privacy is often misinterpreted as distrust and can offend the business partners. When interactions move to the digital space, personal interaction is reduced. In agriculture, this is seen as a bad thing. However, in other contexts, this is a desired outcome. As paper F showed, online interactions are a suitable way to overcome social barriers e.g., in the context of seeking medical help. Many participants stated that they do not want to be seen in a medical office to avoid rumors about them. This is even more critical when stigmatized topics are involved. Furthermore, having acquaintances or family members working in the medical office is a barrier for people who do not want to disclose information on their health issues to those people.

Another important aspect of identifying problems and risks in the context of data disclosure is mental models. Especially Papers D and E showed that the awareness of privacy-related topics and privacy competence affect the attitude towards digital tools and their interaction with them. Privacy incidents that became public raise the awareness and caution of users. Furthermore, education and privacy competence allow users to better estimate privacy risks and make informed decisions. For this to happen, it is necessary to enable the users to enact their rights over their data. Otherwise, the awareness of privacy risks will only lead to resignation. In literature, this phenomenon is known as privacy fatigue [39]. In the study presented in Paper E, some participants stated that big companies already have their data, and thus, caring about privacy would be a waste of time and effort. This leads to a strange form of trust towards these companies: Users believe that companies already have their data, and since nothing bad happened so far, the companies seem to treat the data in a responsible way. It seems that this form of gaslighting helps these users cope with the negative implications of their resignation towards privacy. Paper E also sheds light on other types of trust: real trust and illusionary trust. While investigating the effects of transparency on trust, it was found that transparency helped users to understand security mechanisms better and in consequence, build up trust in the system. On the other hand, the increased understanding led to disappointment: Educating the users destroyed false mental models and the illusionary trust that originated from them. This form of misplaced trust is a real danger for users: By overestimating protection, users can easily disclose data unintentionally.

4.1.3 Ethical Aspects

While investigating the B2B aspects of problems that are related to data disclosure, one aspect dominated the discussion on ethical implications of losing data sovereignty: The fact that agricultural companies may gather operational data from their customers, buy up their land, and use a cheaper workforce to cultivate the land utilizing the data of the previous owners. This practice is described by Fraser [20] and was brought up in the focus groups in Papers A and B. This could have a huge impact on society, not only by disrupting an entire domain but also because of its implications for the resilience of a country. If the SMEs in the agricultural supply chain are replaced by a few big players, these become single points of failure in the food industry. The adoption of digital tools and digitized machinery especially makes agriculture prone to cyber-attacks. This can be utilized by malicious actors to disable huge parts of the critical infrastructure of a country. If the food production is done by a multitude of smaller businesses, it is unlikely that a large-scale attack will stop the food supply entirely.

4.2 Research Focus Develop

The second research focus is DEVELOP, which investigates how data disclosure can be done in a privacy- and data sovereignty-preserving manner. The goal is to answer the second sub-question:

RQ2 **How can developers design tools to mitigate risks in data disclosure scenarios?**

The findings regarding this question revolve around two categories: (1) How is the development of tools affected by laws and regulations, and (2) why it is important to know the potential users and their needs to design tools that prevent unintentional data disclosure and enable users to understand and manage the flow of their data.

4.2.1 Laws and Regulations

It is important for developers to adhere to the laws and regulations of the countries in which their potential customers live. Otherwise, the adoption of the technology might fail, resulting in a huge loss.

New technologies that promise new business models and markets might run into issues due to different regulations across the globe. In the domain of agriculture, farmers are faced with different production standards and environmental protection laws. This leads to different market conditions depending on the region. Technologies that aim toward globalization, such as DTL-based marketplaces, might not be adopted due to different market opportunities in each region. German farmers were especially concerned that they have huge market disadvantages in a global setting due to the higher cost of wages, production standards, and environmental protection, such as limitations on fertilizer. Farmers in other countries can produce more cheaply, and therefore, German farmers are not incentivized to adopt the new technology.

Another problem that could hinder the adoption are technical limitations. As Paper B shows, blockchain technologies have a special problem regarding privacy protection laws such as the *General Data Protection Regulation (GDPR)*[1]: One of their key features is immutability, which conflicts with the demand for data deletion after specific periods. Therefore, developers should choose the technologies wisely to build products that can actually be distributed legally.

4.2.2 Know Your Target Group

In HCI research the incorporation of users is a common practice. This sections highlights how this approach was utilized in this thesis and what findings support the state of the research or even advance it.

Developers should invest in research on their primary target group to ensure that the tools they design meet this target group's needs and thus have higher chances of being adopted into practice. Papers A and B showed that the under-representation of SMEs in technology development leads to adoption barriers. Most technology in the agricultural domain is driven by larger players upstream and downstream of the supply chain, neglecting the perspective of farmers. This leads to inhibitors to adoption since the promised benefits of the new technology are nullified by the overhead in administration. New business models might fail due to the lack of time

[1] https://gdpr.eu/

that is required for them (e.g., constant monitoring of the market in the context of direct marketing in a global marketplace). To tackle these problems, Paper C presents challenges and derives requirements to reduce barriers to technology adoption in this domain: Open standards should be used for data modeling to prevent vendor lock-in effects. Furthermore, solutions should be customizable to meet the different needs of all the diverse farming businesses. These should also support multiple types of devices to ensure that the data is always available and data management is not limited to office work. Making transparent who can access what data is crucial to maintaining data sovereignty and empowering users to make informed decisions. Finally, the data access of others must be documented to ensure liability in case of data misuse.

Similarly, Paper D investigated the requirements for the private user context. Utilizing participatory design to guide the development process is an effective way to find suitable personas and meet the needs of the users. Using personas has proven to be an effective way for measuring the needs of users [40]. Paper D extends this concept with the novel approach of support personas to define the level of support that users need in online interactions. Providing tailored UIs for different user groups is crucial. This can be achieved by incorporating the use of personas into the development process. However, users should still be able to change the UIs based on their preferences since no persona can cover the needs of every human being. Again, making the tools customizable gives users the feeling of empowerment and reduces user frustration. This need for customization is also reported by related work [41]. Incorporating educational measures such as transparency (see Paper E) can help revoke false mental models and build trust in the tool. On the other hand, developers might choose to hide everything from the user to prevent data leakage due to wrong usage. Making the tool a black-box and enforcing security-by-default has its limitations: false mental models are not revoked, so overestimation of protection can also lead to data leakage. Therefore, it might be worth investing in the education of the users and enabling them to make their own decisions. Good UI design can assist this process, as Paper F showed: due to the clear separation of the input fields in the non-customizable version, users had the impression that the data was strictly separated, and that affected their mental models regarding this tool. Likewise, the participants who liked the customizable version praised the ease of use mechanisms and defined the data as sensitive. Again, this shows that providing different UIs increases the overall satisfaction of the users because they can choose the most suitable version for themselves.

When designing the controls of a system, the so-called control paradox [34, 35] should be considered. Researchers observed that increased control options led users to disclose more data. A possible explanation was given by participants of the evaluation in Paper F: By granting them control over their data and the disclosure process, their mental models were affected: Usually, they feel like data is taken away from them if they have to fill out a form. by enabling them to deploy additional security measures and define the recipients of their data more precisely, they had the feeling that they are giving the data away based on their decision. As one participant stated: "Everything I did not encrypt, I chose to not encrypt". This difference in perception is essential for building trust. Instead of data taken away from them, users can decide to give it away. This also affects the attitude of users: When data is taken away from them, they "fight back" by disclosing as little as possible. This could be an explanation of the control paradox: If users have the power to decide, they can evaluate, based on individual information, whether it may be disclosed or not, which could result in them disclosing more data due to a better feeling. However, if they are forced to give away data, they will try to withhold as much as possible.

Finally, ethical aspects should always be considered. Especially Papers E and F showed that transparency and education can increase trust in a system. However, there is also dual-use potential. By misguiding the users or presenting false information, the users can be misled and incentivized to disclose data unintentionally.

4.3 Research Focus Adopt

The third research focus is ADOPT and presents the findings on the adoption of security and privacy-preserving technologies. It answers the third sub-question:

RQ3 **How can users be encouraged to adopt technology in order to protect their own data?**

The findings answering this question can be categorized as external and internal factors.

4.3.1 External Factors

Whether technology is adopted or not can depend on external factors such as regulations, prior investments, subsidies, or the hidden cost of adoption.

Paper A illustrates that the adoption of digital tools can be accelerated if farmers already invested in similar technologies. Furthermore, in some sub-domains (e.g., dairy farming), farmers are required to have safety measures such as smart ventilation systems and power generators. For these farmers, the barriers to adoption are quite low since they have already adopted some technologies and there are a lot of benefits provided by digital tools such as milking robots. Dairy farmers already have to handle digital data for documentation and reporting duties, making them the most relevant target group for tools that manage data disclosure scenarios. One sub-domain that benefits least from digitalization is viticulture. Sparse benefits and few governmental regulations grant them the freedom of choice regarding the adoption of digital tools, making them the least prone to unintended data disclosure.

One major factor for technology adoption are the hidden costs: After the initial purchase, the machinery has to be maintained, has running costs, requires training and education for the personnel, and requires time invested for data management. These costs are one of the main factors why farmers organize themselves in machinery rings to reduce the cost for each farmer by sharing machinery. However, the shared use requires the exchange of operational data, creating another data disclosure scenario.

Another cost-related external factor for technology adoption are subsidies. While providing benefits for the farmers, yet another data disclosure scenario emerges: To get the subsidies, farmers have to report their adherence to strict regulations, making them eligible for governmental funding. For this purpose, data has to be processed and disclosed.

4.3.2 Internal Factors

The internal factors are threefold: (1) Trade-offs that decide whether a technology is adopted, (2) social barriers that prevent adoption, and (3) mental models and why it is important to educate the users.

Trade-offs in literature often revolve around sacrificing usability for security or privacy [42–44]. In the following the insights on trade-offs are presented. As already mentioned, Paper A shows that the perceived benefits have to outweigh the perceived risks in order for farmers to adopt new technologies. Depending on the sub-domain, they are more or less incentivized to adopt additional machinery and technologies. Paper A and Paper C introduce another trade-off: privacy vs. time. Farmers usually have a tight schedule and cannot afford to adopt technologies that require extensive data management and office work. This trade-off was also mentioned in Paper F: While most participants agreed that the additional effort to decide what data is considered sensitive is negligible, some participants feared that the effort for the medical staff could increase due to less data availability. However, comparing papers C and F, it becomes clear that the trade-offs that are necessary depend on the context. While it is a real struggle for farmers to decide whether they want to invest more time in more privacy and data sovereignty, the situation with users contacting a medical office was totally different: Due to the poor availability of online services in most medical offices and old customs such as the use of telephone and the procedure at the front desk, there is barely any trade-off. In fact, the proposed interaction concept increases privacy as well as efficiency for many users. People who cannot contact the medical office within working hours, people with telephone anxiety, or those who have social barriers contacting a medical office, benefit the most. Therefore, the existence of trade-offs, as well as their impact on the adoption of technologies, is highly dependent on the context.

Mental models are not only crucial to building trust and revoking illusionary trust (Paper E), but they can also affect the adoption of technologies. For many security- or privacy-enhancing technologies, it is crucial that users perceive the need to adopt them. Related work shows that user often struggle to understand technical terms [45, 46], raising the need for increased awareness. This requires the correct mental models and education of users to estimate risks correctly. This might sound discouraging for developers of such technologies but it might be worth the effort. If more people can evaluate risks correctly, the customer base increases: In the interviews, it was indicated that customers would pay extra for built-in security features such as attestation.

4.4 Assisting Users in Data Disclosure Scenarios

In the previous sections, three different aspects were discussed, each contributing to one of the sub-research questions. Combining these insights, it is possible to answer the main research question:

RQ **How can users be supported in data disclosure scenarios to make informed decisions and manage their data in a privacy-preserving manner?**

After answering all three sub-questions, it is apparent that each research focus cannot be isolated from the others. This aligns with standards in software development such as human-centered design (ISO 9241–210)[2]. Therefore, the findings of the three different directions can be converted into a life-cycle to be comparable to existing software development processes. The transition IDENTIFY → DEVELOP can be described as the derivation of needs arising from identified problems. Then

[2] https://www.iso.org/standard/77520.html

bringing developed solutions into practice (DEVELOP → ADOPT) sounds like the natural thing to do. However, in the case of privacy-enhancing technologies, real-world adoption requires a well-thought-out strategy. Security vulnerabilities might often be mitigated by automatically rolling out a patch, but privacy solutions often require the involvement of the users, which means gaining their trust. The reasons for new security and privacy solutions are often security breaches, data leaks, or similar incidents. These intimidate the users and raise skepticism. Such feelings can lead to reactance [23] or resignation, which both lead to lacking adoption of security tools and privacy enhancing technologies. In order to accelerate the adoption of these tools, the users and their mental models have to be considered. One promising way to address this challenge is transparency. By educating the users and empowering them to act against security and privacy violations, trust can be gained and the adoption of security and privacy solutions is more likely. To conclude, the adoption process should be considered from the earliest stages of development.

At first glance, the model of IDENTIFY → DEVELOP → ADOPT seems to be sufficient: A problem is identified, then human-centered design is applied and in an iterative development process a solution is created. This is then rolled out into practice to solve the problem. An example for this would be: A cross-site-scripting (XSS) vulnerability is disclosed to a website owner (IDENTIFY). Afterward, the code is modified to escape all inputs (DEVELOP). Finally, the patch is rolled out into live service (ADOPT). This process mitigated the risk of a XSS attack. Therefore, the model of IDENTIFY → DEVELOP → ADOPT covered the entire process. Now, let us consider a privacy-related example: It becomes public that companies collect data from their customers without their knowledge and use it for their own business models (IDENTIFY). To combat these privacy violations, new laws are enacted to prevent companies from gathering data from citizens without consent (DEVELOP). All companies adhere to the new law and inform their users about data collection, and ask for permission to process their data (ADOPT). In an ideal world, the problem would now be solved. However, in reality, the problem was not solved by new laws such as the GDPR. Instead, the problem evolved by utilizing techniques known as "Dark Patterns" [12]. Companies provide cookie banners to adhere to the law, but some use psychological tricks to manipulate users to make them consent to the data collection.

This example shows, that the usual development life-cycles do not cover a fundamental problem: Even by fighting the problem with new tools and user interventions, the adoption of these measures itself can evolve the problem instead of solving it, requiring new and different solutions to be developed. This indicates that even solutions for problems cannot be evaluated in an isolated manner. As a developer of privacy and security features it is crucial to predict possible problems that arise from the adoption of the own tools. In the worst case, good intentions can lead to severe harm for users: Problems might evolve, side channels provide new attack vectors, or the underestimated dual-use potential weaponizes a tool that was meant to help people.

Therefore, the scope of this thesis differs from classical development life-cycles in human-centered design: Instead of depicting the iterative process of development on itself, this process is abstracted to the research focus DEVELOP. Since problems originate from the adoption of new technology and the developed solutions could lead to new problems, the entire process can be seen as a sequence of multiple iterations of IDENTIFY → DEVELOP → ADOPT. The model derived from this is shown in Fig. 4.1. A life-cycle was chosen to represent the iterations of technology development and the possibility of evolving problems. The dotted line that separates the problem from the three research directions represents the border between the real world and the perception of the researcher. The problem exists in the real world and is independent of the existence of a researcher. By identifying it, the researcher becomes aware of it and can thus develop solutions for it. Through the adoption of the solution in practice, the problem can be tackled in the real world.

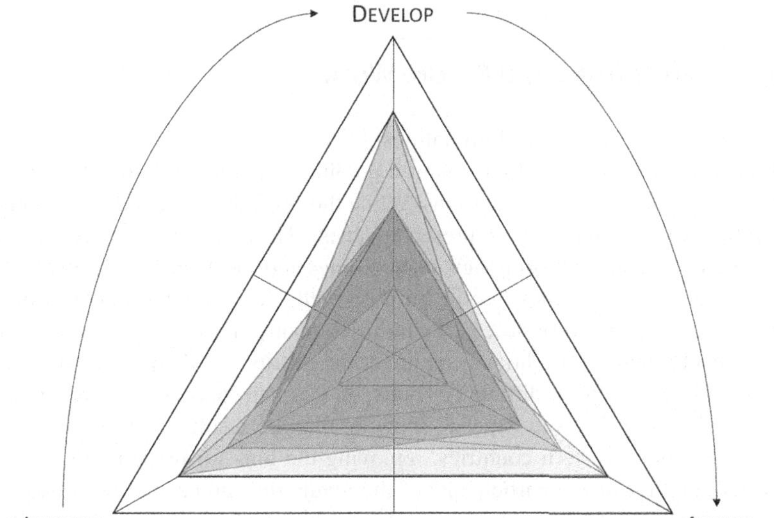

Fig. 4.1 Combination of the three research directions to a combined model

Finally, to compare this model to existing approaches, it is important that this model does not replace the iterative software development life cycle. Both concepts exist parallel with different purposes. Despite both utilizing life cycles to represent themselves, the scope differs significantly. The proposed model in this thesis exceeds far beyond the development of a single technology. In fact, the classical iterative development life-cycle is abstracted within the element DEVELOP: in the scope of this thesis' model, all iterations of development happen in this step and the adoption of the final market-ready product is represented by the transition from DEVELOP to ADOPT. Overall the life cycle depicted din Fig. 4.1 describes the development of an entire domain over time. Why is this relevant for developers? Similar to the general meaning of history it is important to learn from the past to avoid mistakes in the future. if developers learn from the adoption of other technologies and their implications and effects on the original problem, mistakes can be avoided and the adoption of the own tool can be accelerated. To conclude, learning from the past to build a better future is as important for developers as for society overall.

Those who cannot remember the past are condemned to repeat it.
– George de Santayana

4.5 Limitations and Future Work

This work is subject to several limitations:

Firstly, the majority of studies is of a qualitative nature and, therefore, not directly applicable to larger populations. While those studies are good for hypothesis generation, they cannot claim representativeness. However, since the publications within this thesis are exploring new technologies and mechanisms, the exploraitve nature of qualitative studies fits the purpose well. Secondly, the majority of participants were Germans or at least living in Germany. This can be a bias because cultural background or traditions can affect the results and the perspectives of the participants. The cultural differences between Germans and other populations are discussed in more detail in Paper F, showing that the results for Germany can be extended to most western countries, removing the bias for those regions. Lastly, the demographics of the participants of the qualitative studies are not representative. We aimed for diversity in our sample and achieved a gender ratio resembling the actual gender distribution among farm managers in Germany in Papers A and B. Furthermore, related work showed that the ages of our participants in Paper E are similar to the average IoT user. However, matching the demographic criteria of adopters of certain technologies does not exclude a bias: The perspectives of those who are not (yet) adopting technology are under-represented, and, therefore, possible barriers might not be identified. Further demographic limitations revolve around the average high education of the participants and the fact that most of them were from southwest Germany. Especially due to the different historical backgrounds, it might be interesting to interview people from the eastern federal states of Germany. However, as official statistics [18] show, the results are suited for the situation of farmers in western Germany.

To confirm the findings and claims of this thesis, further studies should be conducted. For the development process of the security and privacy-enhancing tools that are presented in the underlying publications, an iterative design process was chosen. This creates the opportunity for further iterations to convert the concepts into usable applications. This way, demonstrators that are only available as demonstrators can be turned into proofs-of-concepts, real-world implementations, and finally, a final product that can be made accessible to the public.

Conclusion

5

This thesis investigates how users can be supported in scenarios where they have to disclose data to third parties. The main research question is:

RQ **How can users be supported in data disclosure scenarios to make informed decisions and manage their data in a privacy-preserving manner?**

To answer this question, three sub-questions are derived, each investigation with a different research focus. The first focus is IDENTIFY and revolves around identifying problems and risks that either originate from data disclosure or can lead to (unintentional) disclosure of data. Secondly, the focus DEVELOP investigates how developers can assist their customers in enabling them to understand their tools and to make informed decisions regarding their data. Finally, the research focus ADOPT sheds light on measures to accelerate the adoption of security- and privacy-preserving technologies into practice, as well as hindrances and barriers to adoption.

From the findings of these three research directions, an overall model is derived to describe the development process for security- and privacy-preserving technologies that enable the user to evaluate risks correctly and make informed decisions regarding their data.

The publications that form the main body of this thesis provide contributions to different areas:

Incorporate users into the design process (Papers A, B, D) HCI research has advocated for a long time that user-centered design is crucial and the development of tailored tools for different purposes is crucial to understand the preferences and needs of different user groups. However, this thesis shows that many perspectives are

S. Linsner, *Privacy Preserving Data Management*, Technology, Peace and Security I Technologie, Frieden und Sicherheit,
https://doi.org/10.1007/978-3-658-47445-4_5

still underrepresented in software development. While it may be profitable for some companies to obtain as much data as possible from their customers, the oversight of their needs and preferences slows down the adoption of technology and digital transformation. Papers A and B emphasize the perspective of those who are underrepresented in the development process in their domain, pointing out reasons for the low adoption rate for digital tools and services. Participatory design was used in Paper D to derive support personas to optimize the development process for online interactions for private users. The novel approach of measuring the need for support of users could advance the field of user-centered development.

Educate the users (Papers D, E, and F) Utilizing transparency interventions, the users are provided with educational measures to help them understand security and privacy mechanisms better and evaluate risks correctly. In this thesis, the connection between understanding and trust was investigated, raising the need for educational measures such as transparency. This helps to build trust in security and privacy mechanisms and reduces illusionary trust, which arises from false expectations and can lead to unintended data disclosure.

Empower the users (Papers C and F) Giving users control over the distribution of their data enables them to enact data sovereignty and protect themselves. This thesis shows that the ability to decide themselves changed the users' attitude towards data sharing: Instead of having their data forcibly taken away, they felt like they were sharing the data willingly based on an own decision. In the qualitative studies of Paper C, D and F the need for customizable tools is highlighted to avoid frustration for users and enable them to disclose data according to their preferences.

Adoption of state-of-the-art research (Paper F) Security and privacy researchers are constantly developing new tools, protocols, and solutions to enhance the lives of users. However, without the adoption of proper use cases, many solutions may remain unused. By exploring everyday use cases for new security features the possible adoption into practice is bridging the gap between academia and the citizens. In this thesis a new interaction concept is presented that utilizes layered encryption combined with deniability used for online forms, which are used in practice by a multitude of actors. With the example of medical offices one of these use cases is investigated. The results showed that users could gain huge benefits regarding privacy and security by the minimal overhead of defining which data should be additionally encrypted. Furthermore, in the current situation, no trade-off between usability and privacy is needed because the proposed solution increased both factors compared to the status quo in Germany.

To conclude, this thesis investigates different approaches to support users in data disclosure scenarios. To accomplish this, different research areas are advanced, confirming results from existing literature and introducing new concepts. Developers of security- and privacy-enhancing tools should consider the history of technology development in their domain to estimate the impact of their product on the problem itself. In order to ensure a successful adoption by users, the tools should address their needs and preferences, while granting them the possibility to customize their experience. To foster the correct use of these tools and to raise the awareness for the need to adopt them, the mental models of users have to be formed in an ethical manner. False attributions should be revoked to avoid unintentional data leakage and real trust can be built.

To conclude, this thesis investigates different approaches to support users in [illegible]

Part II
Publications

Paper A: The Role of Privacy in Digitalization—Analyzing Perspectives of German Farmers

6

Abstract

Technological progress can disrupt domains and change the way we work and collaborate. This paper presents a qualitative study with 52 German farmers that investigates the impact of the ongoing digitalization process in agriculture and discusses the implications for privacy research. As in other domains, the introduction of digital tools and services leads to the data itself becoming a resource. Sharing this data with products along the supply chain is favored by retailers and consumers, who benefit from traceability through transparency. However, transparency can pose a privacy risk. Having insight into the business data of others along the supply chain provides an advantage in terms of market position. This is particularly true in agriculture, where there is already a significant imbalance of power between actors. A multitude of small and medium-sized farming businesses are opposed by large upstream and downstream players that drive technological innovation. Further weakening the market position of farmers could lead to severe consequences for the entire sector. We found that on the one hand, privacy behaviors are affected by adoption of digitalization, and on the other hand, privacy itself influences adoption of digital tools. Our study sheds light on the emerging challenges for farmers and the role of privacy in the process of digitalization in agriculture.

Original Publication *S. Linsner, F. Kuntke, E. Steinbrink, J. Franken, and C. Reuter, "The Role of Privacy in Digitalization—Analysing the German Farmers' Perspective,"* Proceedings on Privacy Enhancing Technologies (PoPETs), *vol. 2021, no. 3, pp. 334–350, 2021*. DOI: https://doi.org/10.2478/popets-2021-0050

S. Linsner, *Privacy Preserving Data Management*, Technology, Peace and Security I Technologie, Frieden und Sicherheit,
https://doi.org/10.1007/978-3-658-47445-4_6

6.1 Introduction

Digitalization in agriculture is a process with very heterogeneous implementations by different actors. This makes this domain an interesting field of investigation in terms of the extent to which certain factors influence the adoption of digitalization. According to Gandorfer, Schleicher, Heuser, and Pfeiffer [47], privacy is a factor that tends to slow the process of digitalization. The processing and exchange of data is a key element of digitalization, but not everyone is in favor of this development. Nevertheless, agriculture is an economic domain that relies heavily on the division of labor and collaboration [48]. Farmers usually cannot grow crops or breed cattle on their own. Multiple actors are involved in the whole process, from planning a season to delivering products to retailers. For these cooperations to function smoothly in times of digitalization, it is necessary to share data in order to be able to plan the individual production steps effectively.

Since data exchange is a major issue, privacy concerns are raised and trade-offs are necessary: High-tech machines can help to save resources and protect the environment, but they require comprehensive and processed data. This data is generated from a variety of information sources and is more useful the more information is available. However, at the same time, this availability can also be a problem: If farmers disclose too much information to their business partners, they run the risk of being put at a competitive disadvantage by individualized prices. Such fears paralyze enthusiasm for digitalization, especially if data flows and purposes are not clearly communicated and contractually secured. To find out how the heterogeneous adoption status and the long duration of the digitalization process in agriculture are influenced by data protection aspects and how digitalization affects the work processes of stakeholders, we conducted an empirical study to answer the following research question:

How does privacy affect the adoption of digital technology in agriculture?

This paper is structured as follows: Sect. 6.2 presents the background and related work of digitalization and privacy in agriculture, as well as the research gap. Building on this, Sect. 6.3 describes our methods, the participants involved in our study, the study design, and the data analysis. Sect. 6.4 presents the results of our empirical study, including the attitudes and concerns of the farmers. Based on our findings, Sect. 6.5 discusses our results with reference to our research question. Sect. 6.6 concludes our study by recapitulating the main findings.

6.2 Background and Related Work

This section provides a brief overview of the context of our study: digitalization in agriculture. Although this paper focuses on privacy for and perceptions of farmers, background information on digitalization in general is helpful to understand the statements of the study participants in this context.

Even though the digitalization of agricultural processes is not a new idea, it is an ongoing development, especially in relation to the current introduction of modern concepts such as IoT or big data in this field [49, 50]. In this context, data privacy appears to be an important factor in the adoption of new digital technology [51–53]. However, studies conclude that digitalization in agriculture is lagging behind expectations [47]. Accordingly, other domains are more advanced in the integration of business models and processes. In the following section, we will summarize recent developments in the field of digitalization in agriculture (6.2.1), also called Smart Farming or Precision Agriculture, to provide context for the reader. We will then focus specifically on privacy and data ownership issues (6.2.2). Futhermore, we will discuss the role of user perceptions in relation to data privacy (6.2.3). The section concludes by identifying a research gap that our study addresses (6.2.4).

6.2.1 Background on Digitalization in Agriculture

Several benefits of digitalized agriculture are mentioned in previous research: First, it improves traceability. Retailers could offer their customers information about the origin of their crops. This could prevent or limit food scandals even more efficiently. One example from research Kamath [54] suggests that better traceability may simplify countermeasures during food contamination scandals. Here, the author refers to two food scandals, in 2006 in the U.S. and 2011 in China, where contaminated products from a single farm damaged the image of the entire sector due to a lack of traceability. In this context, a blockchain-based approach is presented to enable transparency and traceability in agriculture. Similar approaches to this objective exist in further research [55, 56].

Second, digitalized agricultural machinery and equipment could also bring monetary benefits. So-called smart farming approaches promise to increase efficiency and effectiveness [36, 37] through precise maneuvering and application of seeds, fertilizer, and other resources. Taking advantage of these benefits can save time and financial resources. Elijah, Rahman, Orikumhi, Leow, and Hindia [57] also see the benefits of IoT in reducing needed resources while feeding a growing population. Rosskopf and Wagner [58] conducted annual studies from 2002 to 2005 to investigate the usage of computers and electronic devices in German agriculture. Main challenges were lack of understanding of computers and time spent without perceived benefits. In 2017 Gandorfer, Schleicher, Heuser, and Pfeiffer [47] confirmed these findings and stated that privacy is a particularly relevant issue.

Third, the precise application of agents and better calculation based on sensor data could reduce pesticide contamination and thus environmental pollution. As early as 2007, Pinaki and Tewari [59] show in their review of trends in precision farming that there is enormous potential for environmentally sustainable agriculture, an argument which Finger, Swinton, El Benni, and Walter [60] also provide. A meta-study of energy use in precision farming is provided by Pelletier, Audsley, Brodt, Garnett, Henriksson, Kendall, Kramer, Murphy, Nemecek, and Troell [61]. The authors compare different approaches and sub-domains, such as livestock or crop production.

6.2.2 Privacy in Digitalized Agriculture

The adoption of digital tools is closely linked to the handling of data, which makes privacy an important factor. Shepherd, Turner, Small, and Wheeler [62] approach the topic of digitalized agriculture from a socio-ethical perspective: They point out that digitalization in agriculture could help feed the growing population, but success depends on business models that can ensure data privacy and security. The desired increase in agricultural efficiency depends on the establishment of new technologies such as IoT and data analytics in agriculture. The need for security and privacy as well as data ownership is also emphasized by Elijah, Rahman, Orikumhi, Leow, and Hindia [57]. In addition to general security issues in the IoT world, agricultural IoT devices are also vulnerable to physical tampering, such as theft or animals attacks. Looking at the cloud-based backend infrastructure, successful attacks can lead to unauthorized data access. The problem of data privacy is not exclusive to agriculture. Privacy and secure data processing are also important in other areas where IoT is used to prevent de-anonymization or re-identification of individuals [63].

The increasing impact of aggregated and processed data on agriculture is highlighted by Sykuta [64]. The author proclaims principles of big data for agriculture and mentions privacy as an issue. Nery, Santos, Santos, Lourenco, and Moreno [65] name knowledge engineering as a proposed solution. The authors point out challenges such as the semantic gap, dealing with spatial and temporal information, and correlation issues. Fleming, Jakku, Lim-Camacho, Taylor, and Thorburn [66] present perspectives of the industry with a focus on big data. The authors note the need to address issues such as trust, equity, distribution of benefits, or access.

Research also focuses on different countries and their attitudes toward digitalization: Specific drivers for digitalization in Australian agriculture are presented by Zhang, Jakku, Llewellyn, and Bake [67]. The authors interviewed 1,000 Australian farmers from 17 subsectors about their expectations and needs regarding digital agriculture. One striking finding was that the majority of farmers were highly critical of various data assets and wanted more privacy, but were still keen to share data with other stakeholders in agriculture, such as big companies. Fountas, Blackmore, Ess, Hawkins, Blumhoff, Lowenberg-Deboer, and Sorensen [68] asked 198 farmers in the U.S. and Denmark about their attitude towards precision agriculture. They found that the main problem was too time-consuming data handling and that 80% of farmers wanted to store their data themselves. Carbonell [69] sees power asymmetry between farmers and agribusinesses as a problem. The author calls for open source tools and open data for a fairer use of big data. An overview of the adoption of digital tools in agriculture in different EU countries is provided by the study of Kernecker, Knierim, Wurbs, Kraus, and Borges [70]. A total of 287 participants from seven countries participated in this study. It revealed that farmers wanted more instructions and security. It also became clear that most farmers with more than 500 ha land run fully digitalized businesses. Especially the smaller farms still lack digitalization.

The fact that small enterprises in particular lack digitalization has also been concluded by Regan, Green, Maher, *et al.* [71]. As an example, the authors refer to agriculture in Ireland, which consists mostly of family-run farms. They present an interesting view on data ownership and maintaining privacy for farmers. The researchers found a general distrust towards companies, but a very open attitude towards actors with whom the farmers had longstanding partnerships. The authors assume that the reason for this is the family-owned business model. This theory is supported by the work of Cravotta and Grottke [72]. They conducted a study that highlights the tendency of family-run enterprises to favor old-fashioned over innovative solutions. In Germany the demographic situation is similar, as shown by federal statistics [18]: The majority of farmers cultivate less than 200 ha of land. This circumstance makes it worth investigating whether small and medium-sized enterprises (SMEs) in particular are lagging behind in digitalization and whether the lack of viable privacy solutions is a reason for this. For the purpose of this paper, we use the definition of the European Union when referring to SMEs[1].

[1] https://ec.europa.eu/growth/smes/sme-definition_en,
based on headcount (micro $<$ 10, small $<$ 50, medium $<$ 250) and turnover (micro $\leq$ 2 million €, small $\leq$ 10 million €, medium $\leq$ 50 million €) of the enterprises

Furthermore, previous work has outlined that access to corporate data is an existential problem for farmers, as noted by Fraser [20]. Increasing „data grab“ can lead to „land grab“. Once companies have access to the business data, they can easily overtake the farm. By acquiring many smaller farms, companies can manage large scaled agricultural businesses with the data they obtained from former owners. With less effort, the companies are able to gain much more profit from the land than many small farms before. Ferris [52] sees opportunities in precision agriculture, but also dangers arising from the massive collection of data: Exposure of personal data, income, or yield of the fields. The author states that farmers fear disadvantages if this data is accessible for their competitors. Therefore, the author calls for the need for governmental regulation.

6.2.3 The Influence of Users' Perception on Privacy Preferences

The previous section has shown that privacy is an important issue that requires specialized techniques to protect end-user data. However, when developing privacy-enhancing solutions for specific use cases, it is necessary to investigate the behavior and preferences of the target audience. Not only privacy and security behavior [73], but also user perception and reality often differ, as research in other domains shows:

Malkin, Deatrick, Tong, Wijesekera, Egelman, and Wagner [74] investigated the perception of users with regard to smart speakers and found serious misconceptions. About half of the participants were unaware that smart speaker recordings are permanently stored. Furthermore, most users were not familiar with the available privacy functions. 23,8% plan to use them in the future. Users' perceptions of smart home technology were studied by Zimmermann, Gerber, Marky, Böck, and Kirchbuchner [75]. The researchers conducted 42 semi-structured qualitative interviews with inexperienced users of smart home technology and found that users not only fear attacks, but also feel they are losing control. Another example of differences between the mental model of users and real-world technology was found by Han, Reyes, Feal, Reardon, Wijesekera, Vallina-Rodriguez, Elazari, Bamberger, and Egelman [76]. Their study examined the differences between free apps and their paid versions. After assessing which mental model the users had regarding these apps, the researchers found that only 3,7% of the 5877 pairs of apps had significant differences in their use of permissions and data usage. This contradicted most users' impression that paid apps were more privacy protective than free versions. Reasons for specific perceptions regarding privacy and digital tools were investigated by Smullen, Feng, Zhang, and Sadeh [77], who found that users' preferences are related to a specific purpose. Coopamootoo and Groß [78] found that privacy preferences and willingness to share data are based on a person's personality. The researchers identified specific personality traits and their influence on attitude towards privacy. However, these approaches focus on the private individual and their use of technology in their everyday lives. Considering that digitalization affects the business aspects of peoples' lives, it has to be expected that factors other than personality are key to understanding the motivation to adopt or not to adopt. Career implications must also be considered. An approach that considers these perceived negative consequences of online tracking was conducted by Melicher, Sharif, Tan, Bauer, Christodorescu, and Leon [51]. The qualitative interviews showed that users are distrustful of tools in the context of tracking and fear risks such as price discrimination. Although this addressed a monetary factor that influences privacy attitudes, the impact of tracking on individuals is less severe than in a business context.

6.2.4 Research Gap

In the previous sections, we presented the state of research on digitalization in agriculture and the role of privacy. Agriculture relies on the division of labor and therefore the sharing of operational data [48]. Additionally, farmers have reporting obligations to authorities and retailers strive for transparency to provide traceability to their customers. While most of the privacy research mentioned focuses on the issue of privacy in consumer applications, the implications of digitalized tools and data collection in a commercial context need to be considered as well. In a context where the disclosure of proprietary or sensitive data can lead to financial damage or competitive disadvantages for a business, privacy considerations take on great importance. One industry that is particularly vulnerable to these risks is agriculture. This results from multi-actor supply chains and high demands for transparency and traceability from both commercial and governmental sides. While many studies exist on the establishment of digital tools for transparency [54–56], few examine farmers' perception and their roles in the transforming domain of agriculture. While the aforementioned study by Zhang, Jakku, Llewellyn, and Bake [67] provides some insight into the situation in Australia, it also raises new questions: Why are farmers willing to share their data with third parties when they actually consider it critical in principle? Is this related to the specific economic circumstances in Australia, or does it result from a misconception of privacy, as studies in other areas suggest (see [74, 76])? In terms of structural reasons, Kernecker, Knierim, Wurbs, Kraus, and Borges [70] show for Europe that digital tools are less adopted by smaller enterprises. However, their sample for Germany consists mostly of larger farms, while the agricultural sector in Germany is predominantly characterized by small farm structures. Hassan, Reuter, and Bzhalava [79] demonstrate that the decisions of German SMEs to adopt cloud computing are influenced not only by their perceptions of usefulness, security aspects, and the implementation costs, but also by the internal capabilities of an SME. Moreover, the main drivers of technological innovation are larger companies upstream and downstream in the supply chain. Therefore, the incentives for new technologies come from actors with a significantly stronger market position than SME farmers. This creates an imbalance of power and leaves the perspectives of farmers underrepresented. Considering this situation, where SMEs predominate in the middle of the agricultural supply chain, the perceptions of farmers could provide valuable insights into the challenges and barriers to digitalization adoption, and thus privacy attitudes across this sector.

To conclude: To our knowledge, a study with German SMEs in agriculture with a focus on perceived privacy and their experiences with digitalization is currently both entirely lacking and urgently needed. The contribution of this paper is to provide information about farmers' views on the issues of digitalization and privacy. Further, we elaborate how these aspects correspond to the adoption of new technologies. This provides a broad information basis for future studies and allows to address these topics appropriately, taking into account the subjective perspective of farmers.

6.3 Method

Our study aims to find privacy-related issues and obstacles in the adoption of digitalization in agriculture. In the context of this paper, the notion of privacy is not limited to the field of private data, but is extended to the usage for operational data owned by individuals. In this section, we present our overall methodology to address our research question, as mentioned in Sect. 6.1, as well as the design and conduction of the actual study.

6.3.1 Participants

We conducted a qualitative study with 52 participants from agricultural businesses. The study took place at the machinery ring[2] „Maschinen und Betriebshilfsring Rheinhessen-Nahe-Donnersberg"", „John Deere European Technology Innovation Center (ETIC)" and „Hofgut Neumühle", a training and research farm. In prepa-

[2] Machinery rings are associations of farmers which organize collaborative work orders and the use of shared machinery

ration for the actual study, we consulted stakeholders from agriculture, such as farmers, machinery manufacturers, and representatives of farmers' associations, in regular meetings every 2 weeks for more than 8 months, and discussed typical work routines and technological innovations, as well as the challenges of data sharing in agriculture, the parties involved, and regulations. This helped in the preparation of the interview guidelines by pointing out relevant topics and potential conflicts in advance.

We are working in a publicly funded research project called HyServ with partners from the private sector, federal institutions, and associations for farmers, such as machinery rings. Their clients and members were invited to participate in our focus groups. Everyone participated voluntarily and no compensation was paid. Each participant was informed about the objectives and topics of the study via a informed consent form, which was signed by each person. On the advice of our project partners, we launched events for farmers to meet and exchange ideas and expertise or learn about new products and services offered by one of our project partners. In this way, we planned events that were conducted over five days. The first event was a collaboration with the machinery ring, the second with John Deere ETIC. With Hofgut Neumühle we held events on three days due to the high number of participants. During the events, the farmers had the opportunity to attend different program points. One was the focus group interview presented in this study, the second was an agronomy workshop, and the third was a presentation of a NIER-sensor for the analysis of liquid manure. Offering multiple program points increased the motivation to participate by providing a better cost-benefit ratio of travel and offered content. We interviewed the participants in focus groups of 3 to 6 people [30, 31] with a duration of 25 to 30 minutes. These focus groups were conducted by two of our researchers and explored the participants' experiences regarding digitalization and privacy in their systems. I this way, we were able to recruit 52 participants, who own family-run farms in south-western Germany, which can be considered as SMEs. This region is quite rural and has a long history of agriculture. Furthermore, the climate and soil in this region is suitable for viticulture, which allowed us to investigate this particular branch of agriculture.

Our aim was to involve participants at the decision-making levels. Therefore, each participant in our study owns or manages an agricultural business. Most of them run farms, but some also provide services to other farmers. Additionally to farmers, we interviewed one service provider who runs a soil laboratory for farmers and two representatives of the administration, including the head of the local machinery ring and a counselor from a federal administration. For further studies, it would be interesting to approach stakeholders downstream or upstream the supply chain in order to broaden the perspective.

Seven of the participants identified themselves as female and 45 as male, thus the proportion of female participants is 13.5%. According to the 2016 Eurostat database, the overall gender ratio of agricultural workers in Germany was 32.4% female compared to 67.6% male at the date of the census. Nevertheless, as this paper focuses on the operational level of farm managers, the gender ratio of the survey is very similar to the gender ratio of 9.0% for female farm managers in Germany (see [80]).

We recruited most of the participants in the three events with Hofgut Neumühle through a nationwide advanced training institution for agriculture that offers different degrees for farmers after a few years of practical experience. In fact, it is mainly relatively young farm managers who attend this institution to further their education and skills. Therefore, these 42 participants are in the age segment between 20 and 30 years, which is why our study has a focus on the younger generation. However, all these participants grew up on farms and have been familiar with the daily work of a farmer since childhood. The rest of the participants were between 30 and 60 years old.

It should also be mentioned that all of the businesses surveyed were small and medium-sized enterprises. This is because most of the farms in this domain are family-run farms which are inherited over generations. Additionally, according to [71, 72], the adoption of digitalization is a major challenge, especially for SMEs, e.g., in raising equity capital for the adoption. Furthermore, Cravotta and Grottke [72] point to social reasons as challenges for family-run enterprises, such as focusing on owner vision rather than efficiency. This makes these businesses interesting for investigation of the role of privacy in their adoption decisions. For an overview of the fields of work of our participants, see Table 6.1.

Table 6.1 Branches the participants work in (multiple possible)

<table>
<tr><th colspan="3">Branch</th><th>Amount</th></tr>
<tr><td colspan="3">Cultivation of grain</td><td>22</td></tr>
<tr><td colspan="3">Viticulture</td><td>3</td></tr>
<tr><td colspan="3">Cultivation of vegetables</td><td>1</td></tr>
<tr><td rowspan="5">Husbandry</td><td rowspan="2">Beef raising</td><td>Dairy cattle</td><td>12</td></tr>
<tr><td>Breeding</td><td>4</td></tr>
<tr><td colspan="2">Pig housing</td><td>4</td></tr>
<tr><td colspan="2">Laying hens</td><td>3</td></tr>
<tr><td colspan="2">Biogas production</td><td>3</td></tr>
<tr><td colspan="3">Service provider</td><td>6</td></tr>
</table>

6.3.2 Study Design and Ethical Considerations

We interviewed the participants in focus groups [30, 31] because this gave them the opportunity to discuss among themselves as well. In our case, these discussions brought to light new aspects that might have remained undiscovered in individual interviews. All focus groups were led by two researchers to mitigate the likelihood of subjective bias. The entire process, including the creation of an interview guideline, recruitment, conduction of the focus groups, and data analysis and storage followed the guidelines of the Ethics Committee of the Technical University of Darmstadt.

The 52 participants in this study were interviewed in twelve sessions: For the first focus group, we consulted the local machinery ring. This way, it was possible to form an expert panel that included the head of the machinery ring, the soil laboratory owner, and the federal counselor. The aim was to conduct an exemplary focus group interview with them and to review and validate our interview guidelines with these domain experts. The second and third focus groups were conducted with the help of our project partner, John Deere ETIC, who invited customers to participate in the interview. The participants split into two groups, avoiding any (unconscious) bias by manually selecting participants. The remaining nine focus groups were conducted during the three events with Hofgut Neumühle. The participants were farm managers who took part in a federal graduation program to earn the title „state-recognized technician in the field of agriculture (German: Staatlich geprüfte(r) Techniker(in), Fachrichtung Landbau)". Again, the participants were divided into groups to attend the different sessions of the event. In view of the limited time available for the interviews, we decided to outsource some background information into a survey in order to give more room for discussion in the focus groups. The survey was filled out before the focus groups and contained some general information about the branches they work in and their experience with digital tools.

In the focus groups, we asked about their understanding of digitalization, positive and negative aspects, fears, and (if not mentioned by themselves) questions regarding privacy and data ownership. It has to be noted that nearly all groups mentioned privacy aspects on their own initiative. Therefore, we conclude that it is an important issue worth investigating from their perspective as well.

We encouraged the participants to discuss freely about the topics we gave them. Nevertheless, we prepared some questions to give impulses to the discussion, mainly aimed at exploring the perception and state of digitalized agriculture within the focus groups, as the direct question about privacy is susceptible to the acquiescence bias:

- *What is your perception of digitalization in agriculture?*
- *Which digital tools or machines do you use?*

- *Does your farm have its own server or other network infrastructure?*
- *What are your experiences with digitalization in the daily work routine?*

6.3.3 Data Analysis

Data from the focus groups were obtained through audio recordings. Later, these recordings were transcribed and anonymized for coding. We segmented the data into meaningful expressions using the open coding method [81]. We then grouped the codes into categories: digitalization in agriculture, privacy, and data ownership as an important aspect of privacy which was mentioned often. Based on this grouping, we were able to get an overview of all statements on the given topics. This allowed us to derive our results, which are presented in the following section. The categorization was performed by one of the researchers who conducted the focus groups. We decided to do so in order to ensure a homogeneous analysis of the data. To avoid subjective bias, the coding was reviewed by the second researcher who participated in the focus groups. The coding resulting from this process was then presented to the other authors. The recorded interviews are in German, however, we translated the statements as literally as possible into English..

In this paper, we refrain from disclosing the clear names of companies mentioned by the participants in order to guarantee a neutral perspective. These companies can be suppliers of agricultural machinery and equipment, e.g., tractors, irrigation systems, or soil sensors. They may also be contracting firms and suppliers of seeds, fertilizers, or animal feed.

6.4 Results of the Empirical Study

In this section, we present the results of the qualitative study. We derive general aspects of digitalization in agriculture, followed by a presentation of the interviewees' positions on privacy and data ownership in particular. We also present some direct citations of statements that expressed farmers' experiences in a concrete and precise way. In this section, citations refer only to the focus group (*fg*) in which they were mentioned in order to ensure the anonymity of the individuals. The quotes in this paper are numbered (e.g., *Q1*) for further reference in the discussion of the results.

6.4.1 Heterogeneous Levels of Experience and Dependencies on Digital Tools

This study found varying levels of decision-making regarding the adaptation of digital technology. Three branches stand out in terms of benefits and freedom of choice regarding adoption; they are provided as examples to illustrate potential differences: *Cattle farms* that rely heavily on digitalization, *plant farms* that reported benefits of digitalization but do not necessarily need to take advantage of it, and *winery productions* that benefit the least from digitalization and therefore have the least motivation to use digitized tools. Reasons for this are highly heterogeneous levels of available technology to benefit from, the need for technology (e.g. milking robots), or legal requirements, such as animal welfare laws which require every affected farmer to provide emergency generators for ventilation systems or milking robots in order to ensure animal health.

Cattle farms rely on digital solutions, as they cannot maintain their operations without machines and robots. Dairy farms cannot guarantee the welfare of the animals without milking robots (fg8), as dairy cows need to be milked daily, otherwise they suffer from severe pain and poisoning that can lead to death. Without robots, this work would be impossible to accomplished, as it takes 30 minutes to milk a single cow manually. In breeding farms, it is necessary to install intelligent ventilation systems, as the evaporations of the animals would otherwise lead to suffocation. For this reason, farmers who raise animals are under legal obligation to possess generators to keep the machinery running in the event of a power outage.

Since cattle farms rely on the use of state-of-the-art technology, there is little inhibition to adapt to digitalization. Moreover, subjects reported additional optional advantages, such as digital automatic feeding machines, which allocate the optimal amount of nutrients to each individual animal and optimize the performance of the animals (fg11).

Plant production companies are representatives of businesses who do not necessarily rely on the use of high-tech machines and robots, but can rely on a wide range of digitized agricultural machinery and administrative tools. The focus groups in our study also report on the advantages they have experienced through the use of new technologies. These include, for example, agricultural machines that are automatically controlled by satellite signals and can seed fields by making the best use of land and resources. These machines are particularly suitable for angled fields (fg12). This form of precision agriculture is particularly useful in the context of legal requirements such as distance regulations that define zones where agricultural substances may not be used or regulations to protect groundwater and soil quality. In such cases, digitized machines help to apply resources precisely and use them optimally.

Winery productions are least affected by digitalization. Usually, production steps are carried out manually or with non-digitized mechanized equipment, because there are rarely any digitized machines for viticulture (fg2). Only logistics and administration can be optimized by digitalization, but the few advantages are hardly an incentive for winery productions to invest in it. Therefore, some wine producing companies do everything manually or handwritten and even without computers or other machines.

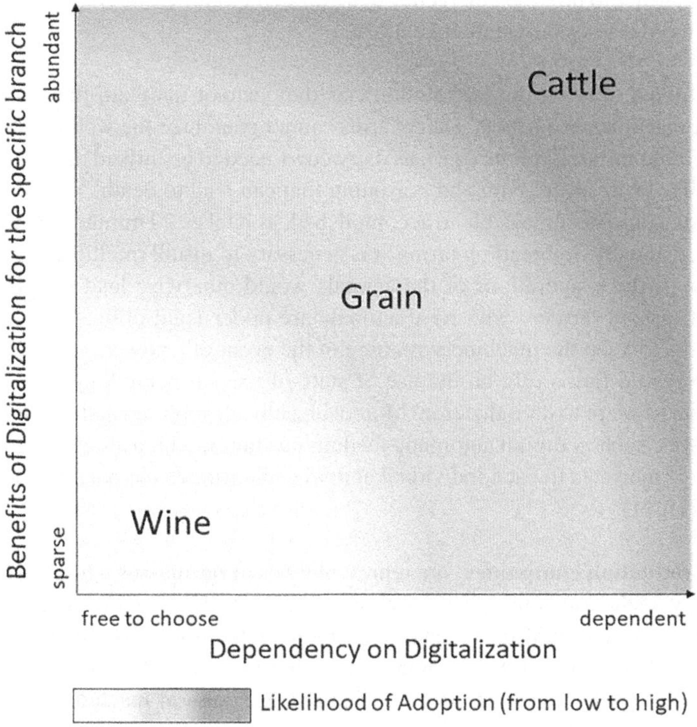

Fig. 6.1 Different agricultural subsectors and how they are affected by digitalization, influencing attitudes toward privacy and the likelihood of adopting new technologies

The fact that farmers differ in the way they are affected by digitalization influences their privacy behavior due to heterogeneous experience or external motivation to expose themselves and their businesses to digital services and privacy risks. In Fig. 6.1, we have illustrated the three prominent agricultural branches of our study in relation to their dependency on digitalization and the benefits of technologies for their subdomain. This figure can be used to estimate the likelihood of adoption and thus exposure to privacy-related technologies of other subdomains.

6.4.2 Attitude towards Digitalization

To provide context for the privacy issues within agriculture, we will present some insights into the general attitudes of farmers towards digitalization in agriculture. Overall, interviewees displayed a balanced view on improvements within their field of work. However, we will elaborate more on the negative aspects, since these are more related to privacy concerns and impede digitalization. Regarding the production steps in the field, automated precision farming is evaluated as helpful for farming within complex field boundaries resulting from the small-scale and fragmented land structure in rural southwestern Germany (fg9), where the group interviews were conducted [82]. Not only is work on the fields affected, but office work is also transforming. Farmers already need to document their work in order to fulfill reporting duties toward authorities. This type of work can be done more efficiently with digitalized farm tools (fg8).

However, not all farmers were generally positive about digitalization and repeatedly mentioned arguments why digitalization is not adopted in all agricultural businesses. One argument is the high price, that is not yet affordable for owners of small and medium-sized enterprises (fg1). Farmers have to decide whether to adopt digital technologies, which ones to adapt and how this will influence their daily work routine (fg9).

> *Q1 (fg10): [...] it is my experience so far that we do not save any work, we just distribute it differently. And perhaps the documentation will then be more centralized, but the skill is to keep the overview*

Another topic that was addressed was that digitalization creates new dependencies. These mainly concerned the technical infrastructure and external services. This refers not only to support for the mechanical components, but also to software failures or mandatory internet connection which causes machines to exit automation mode. Farmers are only left with the option of restarting and hoping that the machine will work again. Otherwise, they have to lay down their work and wait for support services (fg4). The increased dependency on external services results from the need for external staff to fix problems that do not occur with non-digital farm equipment. Another dependency mentioned was availability: Especially the problem of poor mobile data reception in rural areas is a major issue for farmers, because it hinders the effectiveness of their processes on the field (fg10).

6.4.3 The Value of Privacy: What is Being Done with Data in Digitalized Agriculture?

Besides general challenges perceived by farmers, we were especially interested in privacy concerns, which were an important factor for most of the focus groups, based on a lack of trust in the intentions and motives of agricultural companies.

6.4.3.1 The Importance of Privacy for Farmers

The role of farmers is shifting in the process of digitalization. This also affects the different perceptions of the consequences of data transfer towards companies. The scale ranges from fearing the end of the concept of the professional farmer on one side to the opinion that data can be seen as another important resource within the agricultural business on the other side.

> Q2 (fg7): *I see a danger that the farmer might fall behind at some point if the data is really processed in such a way that the companies can take over the planning of the cultivation by themselves.*

Some farmers perceive that they are disclosing more than just data when they give companies access to their documentation:

> Q3 (fg5): *There are people who want our information, how we proceed in the field, or what we have acquired over the years so that someone can analyze it. If they know who did what, where, and when, then almost everyone can copy that. We could be replaced through the many years of experience that we have built up when someone gets this information.*

These statements display the skepticism of some participants that companies are collecting data not only to improve technologies but also to generate fundamental change in agricultural production itself. Experience and lifelong learning in farming, especially in regions with small to mid-scale farms, is perceived as an important factor for effective production. Replacing the factor of human experience with the appliance of accumulated movement patterns and activities could, therefore, lead to increasing automation and gradual de-professionalization of agriculture. Most fundamentally, by giving up the exclusiveness of the knowledge of how to work in a specific field, farmers are at risk of being bought out by companies in the future.

6.4.3.2 Data as a Valuable Resource

Then again, there are also opinions on how to balance the interests of companies in data collection in order to improve technology and the threats that comprehensive transparency poses to farmers. By regarding agricultural data not as a secondary product but as a precious core resource of farming, the prices for data should be commensurate with the advantages companies gain by obtaining data.

> Q4 (fg9): *As long as the data remains within a farm, and I have control over the data, I still see the whole thing [digitalization] relatively relaxed. But as soon as other companies want to gain access to the data, then of course they can also get it at a certain price. So, depending on what kind of data they want, they have to offer something in return.*

From this perspective, there is a need to restrict data access by companies to a certain extent, e.g., only for a short period of time or exclusively for the recipient and not for third parties.

> Q5 (fg6):*The companies that manage the digital crop field cards*[3] *can use the data to create their own personalized profile of you and also predict how you will act in certain situations in the future. That's actually frightening. And who guarantees me that the data will not be sold to other companies?*

Farmers fear that they will be at a market disadvantage if companies can predict their harvest and the effort a farmer has spent in one season. Thus, the perception of data collection is quite negative: if a farmer does not get any benefit from his data, the collection is just additional work with benefits for third parties. This is a serious hindrance to digitalization.

> Q6 (fg10): *We collect a lot of data and do nothing with it. We do collect them, but we cannot use them automatically.*

To conclude, none of the focus groups expressed indifference towards their privacy. Given the perceived risk of professional farmers becoming obsolete and the current

[3] crop field card: administrative tool for planning what measures are applied to which piece of land

situation of data exchange without substantial financial compensation, a need for a solution can be derived by limiting the visibility and accessibility of data by agricultural companies.

6.4.4 Different Actors with Distinct Intentions regarding Data Ownership

As a specific aspect of privacy, many concerns about loss of data ownership were expressed by the participants.

> Q7 (fg5): *Concerning the digital crop field cards*[3] *and their cloud versions too, we all agree on the issue of data ownership. That we reveal a lot about ourselves, a lot goes somewhere unknown or maybe people have access to it, and we don't notice.*

Furthermore, farmers argue that they like to stick with the old-fashioned ways of documentation in order to prevent others from getting insight to operational data:

> Q8 (fg2): *If you write your stuff on paper, you know you have it at home in your office. And if you just type it into a cloud, you don't know who can look in and where the data ends up. That's an unsafe context, because it's about important operational data. I think that is the biggest problem.*

6.4.4.1 Unintended Use of Data

As already mentioned in 6.4.3.1, if companies gain access to agricultural data, farmers run the risk of giving up part of their economic foundations. However, not only economic actors such as big companies or retailers may benefit from data, but also criminal groups or the authorities.

> Q9 (fg6): *I see opportunities in digital agriculture, but I also see risks in digital agriculture in the form of making the whole documentation transparent. That all the data that you collect, that you have on the farm, can or will become public. If data is in clouds, it can fall into the wrong hands, through hackers, for example, and then they can spy on our entire production data, analyze our professional knowledge and then evaluate what we do.*

But hackers are not the only ones who should be kept out: A much more present danger for farmers than ominous hackers are companies. That companies are trying to collect data from farmers is no secret, and some already have business models to share data with third parties:

Q10 (fg4): *So, in some way, you can follow some [digitalization] trends, but you should always have a critical look at them and avoid jumping in headfirst. Because otherwise, you're transparent for the companies. With too much data provided, they can take too much advantage.*

Q11 (fg9): *It is already quite sure that many companies are interested in the data. When I see offers for a digital crop field card*[3]*, where every process on the field is documented, that you get the ten euros cheaper per month, but company "XY" can look into the data. [German Chemistry Company], for example, can look at this data to see what crop protection is being done, what is needed, and for what reason. [...] Not everyone needs to know what I do and what kind of strategy I apply.*

Instead of uploading their data to the servers and clouds of third parties, farmers who already adopted digitalization prefer their own solutions to store and manage their data in order to keep control over it.

Q12 (fg1): *The PC on which the system runs has a security system of its own, then another NAS system is attached to it, then there used to be a cloud backup all the time.*

6.4.4.2 Negative Experiences with Existing Systems

On the other hand, some farmers are using cloud-based systems on their farms, which, however, brings some disadvantages: The fact that many technological innovations in agriculture are developed and offered by the leading companies results in a dependency of farmers on company-specific systems. Another reason that makes farmers dependent on third parties is security. Digitalization is already increasing the office workload, and farmers who want to concentrate on farm work cannot guarantee cyber-security.

Q13 (fg3): *Data protection, also with regard to the security of my data per se, can hardly be guaranteed by myself anymore. I assume that we will soon be looking for a company that will take over the whole thing, where you rent a server, and they take over the data protection part.*

Many responses displayed skepticism about cloud servers offered by companies, which could potentially profit from the data. Indeed, at the same time, reading the privacy policies is perceived as too complex for farmers without in-depth knowledge of privacy law. One participant stated how frustrating it is to be confronted with privacy policies and unintended consequences of accepting them without fully understanding them:

Q14 (fg3): *[Digitalization] is of course a great relief, and you can put an end to all this paperwork, but I don't know where my data will end up. It makes no difference whether I read through their privacy policy or not. Nobody can figure it out anyway. And in the end, there is somehow a [German chemistry company] behind it, which then has my data. And a few weeks later, a letter comes and there are some offers that happen to fit well for my farmland. (approving laughter by other participants) Yes, so you really wonder where this comes from.*

6.4.4.3 Traceability versus Privacy

According to the statements of the participants, newly developed field sensor methods are mostly perceived as convenient and helpful. New technology can provide benefits for the whole supply chain. Being able to trace products from their origin to the end consumer is helpful for marketing, trust building, and avoiding food scandals. Nevertheless, at the same time, this could also provide a loophole for companies like traders or upstream industries to collect data from farmers.

Q15 (fg8): *Just the other week, I have read a report about field sensors from [two German technology companies], which measure all factors like precipitation, nitrogen level, soil compaction, and vegetation. If a company owns this data, they can do everything with it. They can send you your exact fertilization planning. Actually, that is none of their business.*

Technical innovations in the fields of tracking and navigation systems, automated driving, fertilization, irrigation, sowing, and harvest generate large amounts of data, that can be traced back. On the one hand, this is helpful for the farmers themselves, as it simplifies the operational management of a farm. On the other hand, the enhanced traceability of actions through permanent data collection increases the monitorability and accountability of farmers. For this reason, some participants perceive it as a risk that human as well as sensor errors during the whole process of data collection may result in more frequent, unjustified sanctions by the authorities, e.g., for violation of environmental protection rules due to sensor errors.

Q16 (fg8): *A drawback is then, through the accurate data collection, on the one hand, that it is easier for inspectors to retrace activities, but that makes it harder for you to adhere to everything. Just because they see it that way does not mean that it went exactly that way. These are small things like typos or something that can get you into big trouble.*

But not only authorities demand traceability. Customers of farmers, especially in the subsector of organic food grocers, are increasingly requesting traceability in order to serve the demands of the final consumers. Accordingly, the provision of retraceable data serves not only to meet the mandatory requirements of authorities, but can also work as a purchase incentive for customers. At this point, market mechanisms put indirect pressure on farmers to offer more transparency.

> Q17 (fg7): *For retail, I have to provide all my data: when I sprayed [plant protection agents], what I sprayed, when, and what I fertilized. That is what the retail trade wants. In other words, all the encryption [of data] we want for the companies stands in contrast to the traceability that the retail trade wants from us. [...] We have to supply it to the retail trade, because they want traceability, but we don't want to give the data to the plant protection agent companies or [two German fertilizer producers]. But then they get the data from retail trade.*

Offering traceability for customers and consumers does not pose a problem as such for farmers, as it provides only small sets of operational information. At the same time, big retailers may gather a large quantity of data, which could possibly be sold to big agricultural companies (fg5). In this context, some of the focus groups identified a loophole for data leakage towards undesired recipients.

6.5 Discussion

In the following, the results obtained in the study are analyzed and placed in the overall context of privacy in this domain, and it is shown why agricultural SMEs are particularly vulnerable with regard to privacy. First, the impact on farmers is explained and reasons for (or against) adoption are discussed. We then look at the domain as a whole and identify facets that play a role in the adoption of digital tools for agriculture. Subsequently, we elaborate on the conflict between transparency and privacy along the supply chain. Further, we briefly highlight existing approaches from research which could potentially help address the identified problems in the future and place them into the context of our results.

6.5.1 The Impact of Digitalization on Farmers

When talking about privacy, the focus is always on management of digital data and its dissemination or protection. Therefore, privacy relies on digitalization to provide the infrastructure for privacy-relevant services and products. In 6.4.1 and 6.4.2 we presented some information on the impacts of digitalization on farmers. Building on the results of our study, this section analyzes the general impact of digitalization on the domain to provide a contextual basis for the privacy implications.

Technological change affects the work processes of a modern farm in a far-reaching way and changes the profile of professional farmers in the long term. Especially for SMEs this poses a big challenge, since the adoption of automation and digitalization processes are the more profitable the larger the area of tilled land or the number of cattle is. Moreover, the increased workload in the office is problematic for family-run farms with low workforce.

Before digitalization, farmers mainly had to perform manual work. This includes not only work in the field or barn, but also the maintenance of agricultural machinery. Although planning phases and agreements with other stakeholders also existed in the past, as modern agriculture is dependent on a large number of actors and specialized staff, the participants report that the planning and office workload is significantly increased by digitalization. Some even state that the promised reduction in workload due to digitalization is not noticeable, as work simply shifts to the office (see Q1, Q6). The process of data collection for automated machines is perceived as a nuisance by farmers, especially since the benefit for the own business is much smaller (see Q6), compared to the benefits for third parties through the collected data (see 6.5.2). The shift of work to the office requires the farmers, who formerly managed all work steps themselves, to rely on other parties for maintenance or data management. Thereby the risks posed by data propagation are increased even more. This problem of further dependencies also applies to customer retention by the manufacturers of digital tools. Customers may obtain all their digital tools from one supplier only, thus creating a vendor lock-in effect. Furthermore, farmers today are not only dependent on the weather, as they were in the past, but also on good network connections, nationwide mobile communications, and the availability of satellites.

Considering the identified privacy problems of farmers, it is not surprising that the adoption of digitalization is very heterogeneous in this domain. The results presented in Sect. 6.4.1 showed that the subsectors differ regarding the likelihood and the extent of adoption based on the expected benefits and the need to adopt due to market pressure or legal requirements. A higher dependency on the use of technology can have different effects on the privacy behavior: On the one hand, more experience with the technology allows prejudices to be reduced which therefore have less influence on the data management behavior. On the other hand, the dependence on certain technologies can lead to a feeling of being forced to give the data away in any case. Both factors, prior knowledge and resignation, seem to decrease inhibitory factors resulting from privacy concerns. This shows, that privacy is a relevant factor that affects the adoption of digital technology. Another important point, is that in addition to financial and physical resources, farmers have to spend time, share their data, and have to be flexible to take advantages of the benefits of digitalized agriculture.

6.5.2 Conflicts of Interests regarding Privacy and Transparency

Privacy concerns do not only affect the adoption of digital tools, they also play an important role in the everyday life of farmers, since emerging technologies and trends force farmers to provide transparency along the supply chain. The transfer of data along the supply chain is a central feature of digitized processes in agriculture: Data from various parties involved have to be aggregated and exchanged in order to feed machines with the optimal farm data or to guarantee the traceability of certain quality characteristics along the entire supply chain. However, because all the parties involved have their own agendas, conflicts of interests arise.

Many *farmers* consider the demand for transparency as problematic, as they fear competitive disadvantages. While transparency along the supply chain offers advantages for food safety, it weakens the market position of farmers, who have to fear price dumping if retailers know how the season went for each individual (see Q9, Q10, Q11). The constellation of the supply chain, in which producers consist of a large number of SMEs that are supplied by large companies and deliver to large wholesalers and processing companies, is very specific to agriculture. This puts farmers in a weak bargaining position when it comes to protecting their interests. Some of the interviewees reported about tailor-made offers for their farms (see Q14) which placed them in a subordinate position of power. Due to their weakened market position, farmers also have to fear take-over by large corporations, which are able to carry out cost-effective land management by unskilled workers, as they were able to obtain all relevant data from the farmers' experience (see Q2, Q3). Hence, it is difficult for farmers to assess whether the collection and transfer of data for a certain legitimate purpose, such as regulative reporting obligations or more precise services from service providers, may not lead to an unintended disadvantage (see Q17). Therefore, providing transparency and exchanging data is economically not necessarily in the farmers' interests.

However, when discussing about digitalization in agriculture and the impact of privacy (concerns) on the adoption of digital tools, the farmers' perspective is not the only one to be considered. There are several important actors and stakeholders with legitimate data collection intentions which conflict with the before mentioned fears stated by the farmers in the focus groups. Table 6.2 summarizes these conflicts of interests. In the following, we present three prominent stakeholders and their

Table 6.2 Legitimate reasons for transparency and the fears of farmers regarding different actors

	Machine manufacturers	Suppliers and buyers	Government authorities
Legitimate reasons for transparency	→ Better maintenance due to telemetric data	→ Better compliance to quality standards (e.g. 'organic') → Increased food safety	→ Better execution of restrictions and regulations for public safety
Fears of farmers	→ Vendor lock-in effect → Extradiction of agricultural knowledge and experience	→ Price dumping	→ Interference and control by government authorities → Loss of funds

respective interest in receiving data as well as the farmers' concerns with which they are associated.

Agricultural machinery manufacturers: By accessing farmers' operational data and, in particular, telemetry data from machines, agricultural machinery manufacturers can improve their own products and offer optimized maintenance. If a certain part is found to wear out particularly quickly or frequently, farmers' data can be used to better reconstruct and understand the cause and thus optimize maintenance and product design. In this way farmers can hope for better service and warranty evidence, but they must trust that the data is kept and managed securely so that data leakage or deliberate collaboration with third parties do not have a detrimental effect on the farmers. Thus, farmers need to trust business partners, as they have no control over the further use of their data. If this trust is not assured, farmers will refrain from using digitalized services (see 6.4.4.1). This does also apply to cloud services and reflects the fear of loss of control of sensitive business data. The feeling of loss of control is not only limited to the domain of agriculture, as shown by Zimmermann, Gerber, Marky, Böck, and Kirchbuchner [75].

Further tensions arise from producer dependencies: Manufacturers may benefit from high customer loyalty, but farmers thereby become dependent and are not protected from arbitrary pricing. This is the case, because to a certain extent it is more expensive for the farmer to change supplier and convert the entire farm than to pay higher fees from the current supplier. In addition, farmers fear that their farming operations will be recorded and then, after the agronomic knowledge has been appropriated, used to oust the farmers (see Q2, Q3). This results in fears of becoming obsolete, in that companies could gather the experience and knowledge of their profession, take over the farms and use cheap work forces to do the work.

Suppliers and buyers: Farmers depend on suppliers who supply them with seeds, feed, fertilizers, or pesticides. In this area, farmers benefit from transparency, as they can ensure that they receive products that are compliant with the regulations they face, e.g. legal regulations concerning plant protection and fertilization. It is therefore important that farmers document and plan exactly what they apply on their land. Furthermore, this way food scandals could be prevented or detected more efficiently [54] and in the case of food with certain quality characteristics (e.g. "organic") it is easier to prove that the product meets the quality requirements.

In the opposite direction, however, transparency causes problems: If the suppliers know how the farmers cultivate their fields and what they earn from it, the farmers are strongly dependent on the good will of the suppliers. One interviewee reported that suppliers could increase prices for the products in such a way so that hardly any profit remains for the farmers (see Q10). In this way, prices are at a level at which farmers just barely avoid bankruptcy, but at the same time can hardly generate any profit and thus no reserves. A similar problem arises with the buyers: if they get insight into the farm data they can offer individualized, lower prices, which is a enormous disadvantage for the farmers. Our participants stated their own experiences with these offers (see Q14). This problem of unfair competition also endangers existing structures in agriculture. As Linsner, Kuntke, Schmidbauer-Wolf, and Reuter [2] showed, farmers tend to think of business partners as part of their social environment and do not want to give up partnerships that lasted for generations.

Government supervisory authorities: Farmers receive subsidies from public funds, for example from the European Union. In order to receive these subsidies, they must comply with certain conditions such as upper limits for fertilizers, use of certain plant protection products, or distance zones to water bodies in order to protect drinking water. In this regard digitalization would make it easier to document processes on farms, but many farmers feel at the mercy of government control. Also, trust in technology is not very strong. Farmers know from their daily work how susceptible to faults high-tech agricultural machinery is. Therefore, they do not trust that the automatic recording of operational processes by sensors is so precise that they would make their subsidies dependent on it (see Q16).

To conclude, agriculture is a domain in which a large number of small and medium-sized farms depend on large companies to supply them with machines and working materials, offer services such as soil sampling or buy up the yield. Additionally, the farmers are often reliant on IT service providers to digitalize their businesses. Governmental actors need access to data to ensure compliance with regulations. Within this large number of stakeholders, each actor has its own interests and expectations regarding digitalization and data handling. Above all, the transparency and traceability of agricultural products creates tensions in the sector. In order to resolve those, suitable and data protection sensitive processes and tools are needed that take into account the needs of individuals and make the advantages of digitalization available to all.

6.5.3 Outlook on Future Research Possibilities

Concluding our findings on the effect of privacy on digitalized agriculture, we very briefly want to point towards possible measures to address these problems in the future and hence give some impulses for future research.

Our results show that one of the key problems for SMEs lies in their position in the middle of the supply chain and their size resulting in a weaker bargaining position. Sharing too much data towards commercial purchasers or suppliers of necessary primary products and machinery can lead to higher prices and economic pressure. Hence, it is crucial for SMEs to retain the control over the access to and flows of their farm data. This is also important from the perspective of potential (interstate) conflicts [83] or data breaches [84], that may affect farmers. Privacy-enhancing technologies could help to achieve this, by creating usable solutions and access control mechanisms. This could be done with tools based on blockchain-technology, which has the additional benefit of non-repudiation and is favored by the food industry for providing transparency for supply chains [54–56]. However, granting control over data flows for the data owner remains a challenge. For example, if a machinery manufacturer collects data to perform computations on it, even with data access management, data leakage or misuse of data for personalized offers cannot be ruled out. For this purpose, secure multiparty computation methods could enable the manufacturer to perform computations on encrypted or obfuscated data without having access to the actual data. By this, a misuse of data for personalized offers and similar issues could be prevented. While this technology has been suggested for the use in other domains [85, 86], future research could work on creating more possibilities for such an application in agriculture.

Last but not least, we want emphasize the importance of raising awareness for privacy. A situation, in which the sharing of data leads to financial advantages (see 6.4.3.2), could create economic pressure for other producers to share data as well to remain competitive. Hence, in this situation the producers are played off against each other and put under pressure to give up their privacy. As a consequence, greater awareness for these problems could foster the demand for privacy-enhancing technologies and rule out any privacy risks that hinder digitalization in agriculture.

6.5.4 Limitations

Although our study addresses our research question, it still has limitations. (1) Because of the qualitative methodology of this study, the value of it is exploratory and hypothesis-generating. Thus, no quantitative insights can be gained from it and the results may not be applicable to all farms. (2) Moreover, our study focuses on SMEs, because they are the least likely to use digital tools and are therefore worth investigating. However, this is also a limitation, since our findings do not represent every type of agricultural business. (3) While representing the gender ratio of farm managers closely, our study consists of mostly younger participants. A more diverse sample regarding age could offer additional insights.

6.6 Conclusion

In our study, we examined how privacy affects the process of digitalization in agriculture. Such a study might be a valuable background for the research on privacy-enhancing technologies, provided that it presents empirical evidence on privacy-related obstacles and conditions. First, we presented the influence of digitalization on the daily work of farmers and their wishes with regard to privacy (6.5.1). Furthermore, we have shown that different actors along the supply chain have different interests regarding digitalization (6.5.2). Concluding from this analysis, we have presented challenges and possibilities for future development in this domain (6.5.3).

The role of transparency in the industry is controversial. While it offers advantages especially for downstream actors in the supply chain such as retailers, it also creates conflicts of interest for upstream actors such as suppliers who fear being overcharged in price. Asymmetries in market position between SMEs and large agricultural companies seem to amplify these conflicts. For the successful adoption of digitalization without individuals being left behind, it is necessary to establish mechanisms that make relevant data accessible to all without exposing the operational data of individuals for misuse. Therefore, privacy and especially the fear of its violation by new technologies and business practices remains an important factor in the adoption of digitization in agriculture. Many businesses of different sizes have to weigh up whether the promised advantages outweigh the feared disadvantages. Transparency in particular is a double-edged sword: it creates trust, but can also be threatening if business secrets are disclosed to third parties.

Acknowledgments This work was supported by the German Federal Ministry for Education and Research (BMBF) in the project HyServ (01IS17030B), by funds of the German Government's Special Purpose Fund held at Landwirtschaftliche Rentenbank in the project Geobox-II and by the Deutsche For-schungsgemeinschaft (DFG, German Research Foundation)—251805230/GRK 2050.

Paper B: Blockchain in Agriculture 4.0—Empirical Study on Farmers' Expectations Towards Decentralized Services Based on Distributed Ledger Technology

7

Abstract

Cross-industry digitalization, which is also taking place in agriculture through the use of new technology, sensors, and networking, requires the responsible handling of data. One way to manage data and draw value from it is to use blockchain technology. The food industry and end customers, in particular, demand this technology to achieve traceability and transparency. To ensure a meaningful use of blockchain technology for agriculture, this application domain must also be considered and analyzed in terms of its social and operational structures. This paper presents the results of a qualitative study in which 41 stakeholders from the agricultural sector answered questions in a written interview and participated in focus groups. It was found that it is attractive for farmers to adapt to new markets and technologies early on in order to gain a competitive advantage. However, their concerns about losing long-time local business partners and the social fabric in which they grew up must be considered.

Note This paper was published in German and was translated into English for this thesis.

Original Publication *S. Linsner, F. Kuntke, G. M. Schmidbauer-Wolf, and C. Reuter, "Blockchain in der Landwirtschaft 4.0—Empirische Studie zu Erwartungen von Landwirt innen gegenüber dezentralen Services auf Basis von Distributed Ledger Technology," in* Mensch und Computer—Tagungsband, *F. Alt, A. Bulling, and T. Döring, Eds., Hamburg, Germany: ACM, 2019, pp. 103–113.* DOI: https://doi.org/10.1145/3340764.3340799

S. Linsner, *Privacy Preserving Data Management*, Technology, Peace and Security I Technologie, Frieden und Sicherheit,
https://doi.org/10.1007/978-3-658-47445-4_7

7.1 Introduction

The term Agriculture 4.0 refers to the introduction of Industry 4.0 practices into agriculture [87]. This means that people, (agricultural) machinery such as tractors or milking machines, and logistics are linked, and various data on fields, animals, or machine status are collected via multiple sensors in these very devices. At present, farmers already need to process and provide data, for example, to service providers or to fulfill their documentation obligations (e.g., proof of their fertilization activities) to the state. To take full advantage of the machines and their sensors, separate planning is required. If companies rely on Industry 4.0 technologies, they will have great opportunities for optimization [88]. To optimize the current state, it is essential that it is correctly recorded, i.e., that the correct data is and remains available. As a result, virtual goods, i.e., data, are gaining in importance and value. The management of this data presents its own challenges. Particular attention must be paid to the security aspect of data management since agriculture, as part of the food sector, counts as a critical infrastructure in Germany, a so-called KRITIS [89, 90].

Adequate approaches need to be found to provide farmers with fail-safe access to their data and to store that data in a way that preserves the necessary privacy. Decentralization is one approach to establishing fail-safe services in agriculture [91]. One decentralized technology is attracting particular interest in the food industry: distributed ledger technology (DLT), usually implemented in the form of blockchains. DLT is an electronic data storage and processing system in which each participant in a network has a globally agreed-upon status. Unlike centralized approaches, there is no need to trust a central, authorized entity. The global status is generated by the participants of the network with the help of so-called "proofs", which can be witnessed by all participants in a forgery-proof manner. These technologies are currently being tested worldwide to guarantee traceability and transparency in value chains [92]. One motivation for using these technologies is the desire to better combat and uncover food scandals [54], e.g., to be able to check where contamination occurred and to take appropriate consequences—in the short and long term. If these assumptions prove to be true, it is also possible that these technologies could be introduced to the German food and agricultural sector.

It is yet unclear in what form and to what extent this technology will be used in agriculture and which technical and legal requirements must be considered. However, it is already possible to see what fears, expectations, and wishes farmers have of DLT. The resulting research question, which is addressed in this article using qualitative methods, is:

What expectations do farmers have of using DLT on their farms?

This paper is organized as follows: Sect. 7.2 deals with the fundamentals and presents related work. Afterward, Sect. 7.3 outlines and justifies the qualitative methods used, and the results of this analysis are presented in the following Sect. 7.4. Sect. 7.5 discusses possible approaches for DLT-based services in agriculture. Finally, a conclusion is drawn in Sect. 7.6.

7.2 Fundamentals and Related Work

In the following, the stakeholders involved in practical agricultural processes are introduced. Subsequently, the most relevant topic-related scientific publications are presented to derive a research gap.

7.2.1 Stakeholders in Agriculture

Farmers, i.e., the managers of a farm, have to communicate with a wide variety of stakeholders to run their business. One important stakeholder is the *agricultural trader*. Through them, farmers purchase materials such as seed, fertilizer, animal feed, or crop protection products. Moreover, the agricultural trader is usually also the buyer of the harvest. Since there are only a few agricultural traders, depending on the region, farmers usually have only one to supply and sell to. This leads farmers into a situation of dependency on the respective agricultural trader's pricing policy. Other stakeholders include *government agencies* that want to be informed, *consultants* who farmers can voluntarily seek out, and *representatives* who want to sell their products. In addition, there is a multitude of service providers, so-called *contracting companies*, which offer specialized services.

The cooperation of the various stakeholders in agriculture allows processes based on the division of labor and cooperative management models. Large-scale contracts for contracting companies are not uncommon in some sectors. In beet harvesting, for example, large harvesting and logistics chains are used to collectively harvest the land from all farmers in the region and deliver the harvested beets to the processing centers. Initiators of such joint cultivation contracts are not only the large buyers (e.g., Südzucker) but also the machinery ring of the respective region.

Many farmers are organized in these so-called machinery rings. These are registered associations of farmers who, among other things, jointly procure and use machinery or support each other, for example, with labor. Through machinery rings, farmers can offer and purchase both equipment and services: They can arrange for soil samples to be taken from their land, for it to be fertilized, or for their harvest to be outsourced. In the course of digitalization, machinery rings also offered support for less tech-savvy members.

7.2.2 Digitalization and Communication in Agriculture

In four studies from 2002 to 2005, barriers to the acceptance of the advancing digitalization were already worked out [58]. Referencing several studies, the work of Gandorfer et al. [47] shows that although the basic features of digitalization are currently implemented, the use of digital technologies in agriculture is lagging behind expectations. The authors then identified barriers to acceptance through a media analysis of three German-language trade journals between 2008 and 2017. The media analysis shows that obstacles exist not only in the form of high investment requirements and incompatibilities between products from different manufacturers but also, in the recent past, the increased awareness of risks relating to data protection and data sovereignty has led to obstacles to the introduction of new digital tools. Therefore, the authors claim: "*In this context, it seems important in the long term that a transparent market for operational data emerges. Farms need to benefit directly from sharing their data to make Agriculture 4.0 a reality. One component for such a market could be digital data platforms.*" [47]

One area of digitalization that is of particular interest in research is precision agriculture (PA). This describes the management and analysis of small spatial structures in the field and even individual plants rather than looking at entire fields traditionally. In their 2012 study, Aubert et al. [93] examined acceptance criteria specific to PA adoption using 438 questionnaires collected from farmers in the Canadian province of Quebec. The evaluation findings were in line with the expectations that easy use and usefulness are key factors that help novel technologies gain acceptance. The authors highlight that in their study, adoption was found to be independent of the age of the farmers and the size of the land being farmed. Paustian et al. are also engaged in examining PA [94]. In their 2017 study, they examine the adoption of PA by German farmers. Also, they collect demographic data as well as the test subjects' affinity to technology. Using logical regression analysis, they found that there are positive and negative factors of farming circumstances that correlate with positive or negative technology adoption. Positive factors were the presence of service provider contracts, less than 5 or 16–20 years of experience, and a field size of 500 acres or more. Negative correlates were an area of less than 100 ha and the cultivation of barley.

Another important topic for agriculture is communication, as farmers need to communicate with different stakeholders to run their business. These include employees on the farm as well as external stakeholders, such as agricultural traders, customers, and contracting companies. A survey of 168 farm managers conducted by Hobe et al. [95] in late 2017 found that face-to-face conversations are currently preferred as a mode of communication. However, there is also a clearly recognizable trend to communicate with digital communication tools, for example, by using messenger services. This trend from face-to-face conversations towards space- and time-independent communication could strengthen the acceptance of decentralized services such as DLT.

7.2.3 Existing Blockchain Approaches in Agriculture

Blockchain technology is currently very present in the discussion as a basic technology for Industry 4.0, in which physical and virtual systems are to be linked more closely [96]. To achieve this goal, a variety of ways exist to implement blockchain. Maple et al. [97] examine various approaches and compare them with each other in view of relevant criteria such as energy expenditure, performance, scalability, and other features. It is shown that significant differences emerge depending on the implementation and technologies used. For example, an open-access blockchain that uses a "proof of work" has an enormous energy expenditure, which does not apply to a blockchain with a "proof of elapsed time" and authenticated access. This illustrates that not all blockchains are suitable for every use case. In the following, some use cases from agriculture are considered to facilitate a classification of the domain agriculture in the blockchain world:

A 2018 literature study by Bermeo-Almeida et al. [55] lists and analyzes ten major works engaged in studying blockchains in agriculture. The main finding is that most of the works (6 out of 10) deal with the use of blockchains to ensure the traceability of food products or their production process. Furthermore, the importance of blockchain for the Internet of Things (IoT) is highlighted, and privacy solutions are referred to. It is noticeable that the works mainly originate in Asia (7 out of 10)—especially in China (3 out of 10). The authors suggest that the dominance of Chinese research in this area is related to the importance of the agricultural sector in China. Overall, the application of blockchain technology has led to positive outcomes, including (1) transparency of information within the supply chain, (2) addressing IoT issues such as decentralization, anonymity, and security, and (3) improving reliability, fast and efficient operations as well as and scalability.

The use of blockchains in agriculture is assessed less positively by Koens and Poll [98]. The authors examine the basic motivations for its use in three different scenarios in which already implemented blockchain solutions are used. The basic technical suitability of the problem for blockchains is addressed, but philosophical views, (social) network effects, and economic motivations within the application domain are also considered. One of the three scenarios examined is the agricultural production chain. Here, the authors draw on the implementation of Ge et al. [56] in which the different certificates of the supply chain of table grapes from South Africa to Europe can be tracked through a blockchain. Using the scheme of Wüst and Gervais [99], Koens and Poll evaluate the technical suitability of blockchain solutions for a given problem and conclude that there are few technical aspects to justify the use of blockchain in an agricultural scenario. The paper finds that motivations for adopting blockchain solutions are not necessarily technical-rational insights. Social and emotional reasons can also play a role in the adoption of blockchain solutions. For example, potential users are afraid of losing the technical connection if blockchains do not find a way in their domain, and developers are driven by a curiosity to develop appropriate solutions.

In the above-mentioned work, Ge et al. [56] investigate not only the technical solution of the blockchain but also the stakeholders involved and their attitude towards blockchain technologies. A rough categorization of the stakeholders involved, which is unfortunately not methodically presented, introduces four categories that differ in terms of know-how and willingness to use blockchains: blockchain aspirants (low know-how, high willingness), blockchain evangelists (high know-how, high willingness), blockchain skeptics (low know-how, low willingness), and blockchain observers (high know-how, low willingness). The biggest concerns of the stakeholders are the following five points: (1) the correctness of the data stored in the blockchain, (2) environmental burdens due to the necessary computing capacity, (3) how to deal with different future blockchains, (4) the validity and correctness of smart contracts, and (5) the effectiveness of blockchains to prevent fraud in the food chain.

According to the authors, the last point (5) cannot be solved because a blockchain does not prevent someone from distributing *false* data. However, in such a case, the detection and analysis of fraud could be simplified.

Another motivation for the use of blockchain technology is provided in the work of Kamath [54]. In this work, a project by Walmart and IBM is presented in which blockchain technology is used to track food production chains. This was trialed using the production and supply chain of pork and mango as examples. The aim of the initiative is to efficiently track the cause of food-related epidemics while preventing food scandals. The results of the pilot studies show that with the help of blockchain technology, the time required to trace the origin of relevant food products is reduced from several days to a few seconds. This way, the risk of contaminated food can be contained more efficiently, as it can be withdrawn from circulation in a more selective manner. This, in turn, can prevent life-threatening food poisoning. Before the use of blockchain, a cessation of sales had to be imposed on all similar products. For example, this was the case in a 2017 salmonella scandal in the U.S. where toxic papayas were in circulation. Finding the source—in this case, a single farm—had taken several weeks. In the meantime, various uninvolved farms had also suffered financial losses due to the complete secession of the sale of papayas.

7.2.4 Research Gap

As shown in this chapter, a variety of stakeholders exist in agriculture whose business processes would have to be redesigned if blockchain technology is used. Various works [54–56] already indicate that useful solutions for a wide range of problems in agriculture can be found by means of using blockchain technology.

Although DLT is not the only technical way to implement decentralized services, it is receiving a lot of attention from industry and researchers. Therefore, we see a given probability for blockchain technologies to enter the German agricultural sector. Meanwhile, it remains unclear how farmers view such technologies and how to design novel solutions so that they integrate meaningfully into business processes for farmers.

7.3 Method

This paper examines what farmers think of DLT and blockchain technologies, how they accept them, and what concerns—which, if known, could be addressed ahead of time—they have about these technologies. For this purpose, a qualitative study in the form of a written interview and focus groups with 41 farmers was conducted in cooperation with the rural service center of Rheinhessen-Nahe-Hunsrück (Dienstleistungszentrum Ländlicher Raum Rheinhessen-Nahe-Hunsrück)

The age of the participants ranged from 22 to 26 years. Of these, seven were female, and 34 were male. All participants attended a vocational school and a technical school for two years. The study was carried out voluntarily as part of the training "State-Certified Technician, Specialization: Agriculture" (Staatlich geprüfte(r) Techniker(in)), Fachrichtung Landbau) and was carried out at the Training and Research Institute Hofgut Neumühle.

7.3.1 Study Design

The subjects were divided into groups of 4 to 6 participants each. At the beginning of the study, a low-threshold introduction to blockchain technology was given in the form of a narrative. This was devised by an expert who is employed as an agricultural advisor for farmers at the rural service center of Rheinhessen-Nahe-Hunsrück and administrates teaching activities as part of continuing education:

> Formerly, business was conducted directly between a buyer and a seller as a barter transaction with a handshake. The use of money as a means of payment equalized this relationship. [...] Later, cash was no longer necessary[...]. Through the Internet, there is nowadays the possibility of making "handshake business" with trading partners around the world. However, the anonymity of the Internet causes many farmers to worry that the goods will not be paid for correctly [...]. To protect against fraud [in the past], witnesses at the marketplace had to vouch for the trade. With the help of "distributed ledger technology," a "marketplace" can also be simulated on the Internet. [...] The blockchain ensures that each participant at the virtual "marketplace" receives a transcript of all transactions and stores it in his or her private ledger. This way, manipulations can be uncovered as other participants can testify to the original trade.

In the following, the participants were shown a case study from the USA that implements a use case (direct marketing for farmers). Through this, the participants knew that there are already possibilities for the use of DLT in agriculture. After this introduction, the participants were asked to answer the following question in writing:

> Would you rather welcome or reject the introduction of blockchain-based "direct marketing" of grain in Germany? Please give detailed reasons for your opinion.

In this question, the term "blockchain" was chosen because it can seem more comprehensible than DLT. This was to collect their opinion on the use of these technologies in agriculture. Answering questions in writing, i.e., interviews, has the advantage that each participant can write down their thoughts undisturbed. Also, conducting this type of interview involves less working time to collect the data since there is no 1:1 supervision of the respondents. One disadvantage is that it is not possible to ask questions or make comments directly when answering the question, for example, if there is any ambiguity. Nonetheless, interviews are an explorative method of data collection, which is appropriate in the context since the initial aim is to find out how farmers currently think [30].

Subsequently, focus groups [100, 101] with a length of 20–30 minutes were conducted with the participants in their groups. For this purpose, the participants were seated at a table and accompanied by two moderators who conducted the study on-site. In the beginning, participants were asked about their experience with digitalization in agriculture. Other suggested subject areas to discuss were “resilience” and “blockchain”. In each subject area, deeper discussions were actively encouraged by the moderators. Focus groups, like interviews, are also considered a proven way to gain a deeper understanding of the thoughts of the target audience at the beginning of a research project [30]. Hence, conducting focus groups is a useful method for assessing the acceptance of a technology and obtaining information regarding the target group’s concerns and wishes, which should be considered in a future implementation.

7.3.2 Data Collection und Analysis

Since the interview question posed was answered in writing, and this was done by handwriting on-site for technical reasons, the results of this part were already available. Audio recordings were made of the focus groups. The participants agreed to this, both in writing in the form of an informed consent and during the audio recording. The written statements of the results of the interviews were manually prepared in tabular form. The statements on specific subject areas (y-dimension) were collected and evaluated according to person (x-dimension) and color-coded according to emotion (positive, negative, neutral). The subject areas of the statements are presented in the following chapter. The focus group interviews were transcribed and evaluated according to these subject areas.

7.3.3 Limitations

The validity of the study is subject to the following limitations: Due to the fact that farmers who are currently pursuing their technician degree were surveyed, only individuals between the ages of 22 and 26 were interviewed. This age group does not represent the majority of German farmers, as this age group represented less than 10 percent of farmers in 2013 [102]. The biggest age group by volume (more than half) was between 35 and 55 in 2013. It stands to reason that today’s distribution is similar. This is mitigated by the fact that all participants came from family farms. Moreover, they saw themselves as representatives of their older family members and reported on their experiences and concerns.

A second limitation relates to the sectors represented. The interviewed farmers mostly have dairy farms (with and without cheese dairies) and/or produce grain. In addition, some of the respondents run a contracting business. Representatives from other branches of agriculture, such as viticulture, are therefore underrepresented in this survey.

The question arises to what extent the mental model of the interviewed farmers about DLT or blockchain corresponds to reality and to what extent this factor is important. Since this study focuses on the expectations of the respondents, it is more important at this point to capture these very fears and expectations, regardless of whether the mental model of the surveyed farmers of these technologies is correct.

The study design should be neutral towards the discussed technologies DLT and blockchain, but it can be argued that more positive aspects were highlighted and not enough negative ones mentioned. Thus, in further studies, more attention should be paid to ensuring that the study design is not suggestive.

7.4 Results

The farmers' opinions were grouped into the following subject areas: 7.4.1 Direct marketing and prospect of profit, 7.4.2 cooperation and collaborative farming, 7.4.3 effort and usability, 7.4.4 data protection, 7.4.5 transparency, 7.4.6 competition, and 7.4.7 acceptance. The individual subject areas are examined in more detail below. Here, either the number of people who made a statement or a combination of characters in parentheses is listed as an anonymizing ID for a person.

7.4.1 Direct Marketing and Prospect of Profit

The majority of the farmers surveyed (31) recognize advantages in direct marketing via the Internet. They hope for greater profits since the margins of intermediaries are eliminated (15). In addition, they would have a variety of offers to choose from, which is seen positively, especially by those who have only one buyer in their region. S5P2, for example, explicitly expressed that farmers had only one or two agricultural traders in their region. Some of the respondents (9) stated that they do not have the necessary storage capacity to choose the optimal buyer at the right time because they transport their harvest directly from the field to the agricultural trader. Only two explicitly stated that they had their own on-farm storage facilities.

However, farmers see the greatest advantage in the fact that high quality can be remunerated accordingly (11) since they would not receive a standard price from the agricultural traders, as has been the case up to now.

However, there are also critical voices. Eleven farmers expressed the following criticism: Due to the risks in agricultural production, the quality cannot clearly be determined in advance. The farmers fear that they will not be able to sell the goods if they do not meet the customers' requirements, for example, in the case of weather-related quality losses. They doubt the true increase in profits through direct sales and fear that farmers could undercut each other in the fight for customers (4). The feared price fluctuations (S7P4) have a deterrent effect. This process is seen critically by the interviewees, especially in the context of globalization: Although they could trade on the world market, they fear that international differences regarding production and quality standards could displace German farmers (6).

Another point of criticism relates to the lack of affinity for technology among many older farmers: Four respondents said they could lose their farms or be forced to use additional services in order to use such digital services. These expenses would reduce the profits hoped for.

7.4.2 Cooperation and Collaborative Farming

Another perspective is collaborative farming: several farmers could join forces and work on orders that would be impossible for individuals to handle. This further increases the number of potential buyers. However, farmers tend to be skeptical of this concept. The role of their own farm is considered "too insignificant" (19). Three of the interviewees even extended this statement to all farms in their region. They see themselves exposed to the danger of being bought out or forced out of the market (4). Some see a chance for such cooperative approaches only if a machinery ring or a producer group takes over the organization (3).

7.4.3 Effort and Usability

A major problem for digital services is the additional workload, according to the respondents (6). Farmers, especially those who run a small family business, are very busy, they say (S4P3). A service imposing more work on them would be difficult to implement: One participant (S4P3) stated that this would leave her with no time for "real work". Four farmers also fear long familiarization phases and the already described lack of affinity for technology [47, 58]. Especially among the older ones, this causes concern, as the following quote illustrates: "Smaller farms are often managed by older farmers. They are not familiar with digitalization" (S4P1).

In the case of direct marketing, three participants stated that constant market observation to select the optimal offer would exceed their capacities. But even services that do not require constant monitoring and attention must be "user-friendly and well thought through" (S6P3) in order to be successful.

7.4.4 Privacy

The opinion of the respondents on the data protection of the technologies discussed was ambivalent: Two expressed full confidence in the technology and cryptography, but eleven explicitly expressed major reservations. One farmer (S4P2) expressed concern that the sensors were too inaccurate to be used for documentation and monitoring. The damage in case of a malfunction would be too serious (S1P2).

In addition, farmers fear that long-term disadvantages may arise for their profession: One farmer stated that using digital services would make him dependent on a manufacturer (S1P2). This gives rise to further fears: For example, by using cloud services, the entire documentation and planning of a farm would become "transparent" (S6P2 and others). Another farmer (S9S1) expressed his willingness to pay more for services if it were ensured that the product or service is not subsidized by companies that are allowed to analyze customer data in return. In general, the opinion prevails that operational data would be diverted from its intended use or resold. This refers not only to "hacker attacks" (S6P2) but also to the perceived "loss of control" when farmers have to store their data on company servers, which is confirmed by the following statements:

> "Depending on the cloud in which I have stored the things, there is a lot of trading going on, which is not always to our advantage." (S2P4)
>
> "But I don't know where my data will ultimately go to. [...] In the end, there is somehow a [name of a chemical company] behind it, which then has my data, and a few weeks later, the mail comes including a few offers, which quite by chance fit my company data." (S3P1)
>
> "With too much data, they can just do too much." (S4P2)

One farmer (S5P2) even expressed the fear that the farming techniques that have been passed down in the family for generations could be used by large companies to eliminate the profession of the farmer in the long term. Farms would be bought up and managed by unqualified workers who receive instructions via the Internet: "The empirical values are lost that have been acquired over many generations. This is simply given into external hands [...]. That is not necessarily beneficial. At some point, this will lead to the complete abolition of farmers and [...] large corporations will simply dictate the activities without anyone having to think about it [...]" (S5P4, confirmed by S5P1 and S7P1).

7.4.5 Transparency

One motivation for the introduction of DLT in agriculture is traceability, which the interviewees also confirmed. For example, one farmer states: "It maintains the identity of the grain" (S2P4). In addition, transparency promotes the billing of services since the framework parameters are precisely recorded and stored. On the other hand, there was concern that transparency would lead to consequences due to improved governmental control capabilities (S7P6). In the context of the Fertilizer Ordinance (Düngeverodnung—DüV) introduced in 2017, the word "transparency" seems threatening. Farmers do not want to jeopardize their subsidies through unavoidable small deviations (S7P3).

Furthermore, the respondents fear that transparency could jeopardize data protection: "We have to provide it [the data] to the retail because they want to have traceability, but we don't want to give the data to the pesticide companies or [names of companies in the agricultural trade and the pharmaceutical and chemical industry]. But then they get it from the retail" (S7P6). The reason given for hiding the data from suppliers and buyers was the fear that they might personalize prices to get the highest possible price without customers refraining from the service. For example, S8P1 commented, "If I [...] can access data, I can tell you exactly what he earns on that hectare [...]. Then my product is allowed to cost so much that it is profitable for him. So he can produce for [a price of] 100€, but he knows that for 150€ he will somehow get it financed. So I do it for 150€ [...]. I think, especially with pesticides, this is problematic."

7.4.6 Competition

Contrary to previous statements about maximizing profits by omitting the intermediary, most farmers are concerned about the agricultural traders' livelihood (11). Many of them value the convenient service and proximity to customers (10). One farmer (S6P4) reported: "I can call until after 7 p.m., and the next morning, everything I ordered is at the door." Business relationships exist over several generations and are part of the social fabric of farmers. Four interviewees confirmed this in their statements: They appreciated the long-time personal contact directly on-site. For them, conducting business is part of maintaining social contacts, and they fear that these interpersonal relationships could suffer. One farmer even expressed the wish that agricultural traders should be involved in the design of digital services. However, some see the competitive pressure on agricultural traders as positive: By losing their monopoly, they could be forced to offer farmers better prices for the crop yield. This could increase farmers' profit margins (7). Others fear the opposite: To compensate for losses, agricultural traders would raise prices for seeds, animal feed, fertilizers, and pesticides. This would leave farmers with less profit (4).

7.4.7 Acceptance

From the farmers' statements, a range of opinions can be derived. One farmer rejects digitalization as a whole and said: "I am generally in favor of getting away from digitalization a bit and emphasizing our own skills again"(S4P3). However, such disapproval was not present among the other participants. Almost half of the respondents said they would like to try DLT-based services such as those described in the beginning (19). Less than a third do not consider such services useful for their own business (13). The main reason for this is that the size of their own farm is too small (see 7.4.2). It was pointed out that farms in eastern Germany would be more interested, as they have to manage larger areas on average. The overall opinion is shown in Fig. 7.1.

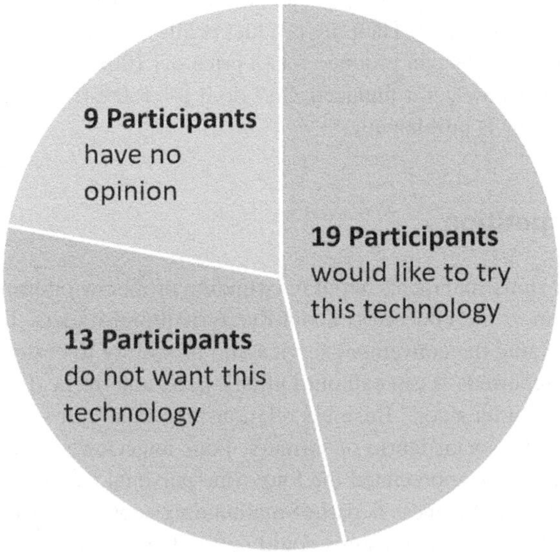

Fig. 7.1 Acceptance of DLT-based services by the farmers surveyed

7.5 Discussion

In the following, the findings from the previous Sect. 7.4 are used to discuss the role of DLT-based services in agriculture.

7.5.1 Reflection on the Results in the Context of Related Work

With this study, some statements of related works from Sect. 7.2 could be confirmed. Rosskopf et al. [58] stated that over the duration of their studies, the use of computers has become more widespread. This trend is confirmed by the above-mentioned results. The farmers interviewed indicated that some older farmers have problems using digital services. However, according to their statements, they themselves can handle computers well. Furthermore, the lack of training found by Rosskopf et al. [58] could be confirmed. Farmers stated that they do not have the necessary time to deal with complicated services. In addition, the preference for personal contact, already reported by Hobe et al. [95], could also be confirmed.

However, other assumptions from related work could not be confirmed: Aubert et al. [93] indicated that the acceptance of new technologies was independent of the farm's size. However, many respondents in our study stated that the limited size of their own business was an obstacle to trying out new technology, as Paustian et al. [94] also found. Small farms cannot afford to follow every trend and invest in unsophisticated technology. The investment is only worthwhile when the technology has reached market maturity and offers reliable added value. These results also support the findings of Gandorfer et al. [47]. Another finding in agreement with Gandorfer et al. is the strong desire of farmers for data protection. Many farmers view their data as business secrets. DLT could be used to create value creation models that allow farmers to share the data they collect themselves in a way that is worthwhile. The application examples motivated by the food industry [54, 56] are already known to the farmers interviewed, as presented in 7.4.5. Below, we discuss how the concerns expressed by farmers can be addressed.

7.5.2 Data Protection in DLT

A key requirement of the farmers surveyed for using DLT-based services is data protection. Almost all respondents expressed a strong need for data protection. The data collected by sensors and the planning are their business secrets and factors for the success of their farm. They do not want to share this data, which is valuable to them, with everyone. Distrust of centralized cloud solutions was explicitly expressed. The trust needed in such a centralized infrastructure's operators is absent. For this reason, decentralized models such as DLT are promising, and it needs to be explored how blockchains and smart contracts can be compatible with the new General Data Protection Regulation (GDPR).

In this context, the immutability and decentralization of the blockchain are problematic. The GDPR requires that data can be deleted and that there is an entity responsible for it. Yet, a general recommendation for such a responsible entity cannot be given, as this is always use-case specific.

However, technical solutions can be found for the right to erasure (Article 17(1) GDPR). Since the term "erasure" is not defined in the GDPR, it is open to legal interpretation. Accordingly, instead of removing a dataset, the anonymization of that dataset could also be considered. In the following, possible technical solutions are presented that would be GDPR-compliant:

One possibility is to store only encrypted data on the blockchain. In the event of an erasure request, the corresponding key would be destroyed. However, this approach also has disadvantages: Encryption methods that are considered secure today could be cracked in the future. In this case, the data could be recovered by unauthorized persons [103]. Therefore, this approach does not meet the requirements.

A better option is the "off-chain" principle: Here, the personal data is not stored in the blockchain itself but outside of it. All that is stored in the blockchain is a hash value that links to the data. The hash itself falls under personal data, but if the data is deleted outside the blockchain (off-chain), nothing is referenced by the hash [103]. This way, the integrity of the blockchain is maintained without revealing confidential data. The storage could be carried out by the farmers themselves or by a responsible body if one exists. The added value of such a blockchain lies in its protection against forgery. In the event of a food scandal, farmers' data can be requested. The checksums in the blockchain ensure that subsequent changes can be detected immediately.

7.5.3 Possible Use Cases for DLT in Agriculture

In addition to the transparent traceability of agricultural goods, other use cases for DLT-based services in agriculture are possible, as the empirical results show. For example, new possibilities are emerging for defining how data is handled. For farmers, the processing of data would thus no longer be a necessary evil brought about by digitalization but an opportunity to create value. The example of weather data illustrates this: Often, farmers operate their own weather stations to determine the optimal weather conditions for work in the field. Using DLT, detailed agreements could be made on the use of this data. For this, smart contracts are a good option. These are self-executing programs built on blockchain technology that respond to the occurrence of various predetermined events. For example, two farmers could agree to share their weather station data with each other. A smart contract could coordinate the sharing of the data so that both receive the other's data once both have submitted the data. For the use case of smart contracts, an embedding in an application with high usability would have to be developed. This is because, so far, these are very complex processes that are not manageable for end users without computer science knowledge.

Such smart contracts could also be used to establish decentralized trading platforms that allow farmers to sell their produce to interested parties without being tied to the limited choice of land trading outlets in their region. However, social aspects are relevant for the design of these services: Farmers highly value established business relationships. While respondents would like more choices in selling their goods, they do not want to lose previous partners. A decentralized trading platform should, therefore, consider the regionality of trade. In addition, laypeople need to be empowered to operate these smart contracts. This challenge exists not only in agriculture but equally in all industries that use smart contracts.

7.6 Conclusion

This paper presents the results of a qualitative study with farmers, in which their expectations towards DLT and the resulting business processes were recorded. Related studies show that there is great interest in DLT on the part of the industry, as it enables the traceability of products [54, 56]. In addition, a faster and more efficient response to food scandals is expected [54]. In order to assess the acceptance of this technology in the event of a market launch in Germany, a qualitative study was conducted with 41 farmers to answer the following research question: "What expectations do farmers have of using DLT on their farms?"

The majority of respondents were open to the possibility of direct marketing via decentralized platforms. However, the concept of collaborative farming was perceived as unattractive. In general, digital services were perceived to be additional expenses. Particularly strong were the farmers' fears that increased transparency in the production chain could also reveal trade secrets and thus strengthen the position of large agricultural companies, which the respondents perceived as a risk to their livelihood. Some farmers saw the impact on current business relationships as problematic. They fear the loss of long-time partners and social contacts. Nevertheless, the proportion of farmers who would try DLT-based services outweighed those who reject them (comp. 7.4.7)

When designing DLT-based services, attention should be paid to sufficiently address the concerns of those affected to improve the acceptance of the new services. If this is ensured, further opportunities for the technology can be exploited. Decentralized services, such as enabling direct marketing via smart contracts, could not only solve the trust issues against cloud services but also offer new value-creation opportunities for farmers.

It is to be noted that the group of farmers surveyed in this study is not representative in terms of their age. Moreover, the level of knowledge of the respondents regarding the concept of "blockchain" and whether DLT or blockchain are the solution to the current demands and problems was not examined. To verify and extend the results of this study, further studies with farmers could be conducted in the future. For example, the attitudes of additional age groups could be surveyed. Quantitative studies regarding the general acceptance of technology by the target group (e.g., with technology acceptance models) and the acceptance of DLT, in particular, are also imaginable. It could also be interesting to investigate in what form DLT can be beneficially applied in (existing) business relationships. Based on this, prototypes (e.g., applications or use cases) could be developed and evaluated with all stakeholders and agricultural actors (i.e., also machinery rings, agricultural traders, and contracting companies). This could be used to research the specific requirements of the target group, e.g., also with regard to usability or other technical aspects, such as the classification established by Maple et al. [97]. Feasibility analyses of different granularity for different use cases would also serve the future discussion.

Acknowledgments This work was supported by the German Federal Ministry of Education and Research (BMBF) within the framework of HyServ (01IS17030B), by funds of the special-purpose fund of the federal government at the Landwirtschaftliche Rentenbank within the GeoBox-I project, by funds of the DFG within the SFB 1053 MAKI as well as the SFB 1119 CROSSING, and by the BMBF and from the Hessian Ministry of Science and the Arts (HMWK) as part of their joint funding for the National Research Center for Applied Cybersecurity. We would like to thank T. Orth, S. Steck, and L. Teichmann, whose student work contributed to Sect. 7.5.2.

Paper C: Supporting Users in Data Disclosure Scenarios in Agriculture through Transparency

8

Abstract

Business collaboration in the era of digital transformation requires the exchange of operational data. Since data are hardly controllable once they have been published or shared with others, it is highly important that users are clearly informed about who has access to which data and how certain settings can prevent the disclosure of sensitive data. However, giving end users more control over their data through increased transparency could also lead to information overload. This is particularly true in the field of agriculture, where tight schedules put pressure on employees of small enterprises. We conduct an empirical prestudy with 52 German farmers to investigate current data sharing scenarios. From these insights, we derive requirements and a concept for data sharing solutions providing data flow transparency for users. To investigate the behavior of users and the effects of transparent UI controls, we evaluate a prototype with 18 persons. Our evaluation shows that farmers demand flexible and secure tools that adjust to their workflows. Also, data should be stored and processed locally, granting farmers data sovereignty. Although the controls require additional effort, the evaluated transparent controls for data disclosure are easy to use and raise user awareness.

Original Publication *S. Linsner, E. Steinbrink, F. Kuntke, J. Franken, and C. Reuter, "Supporting Users in Data Disclosure Scenarios in Agriculture through Transparency,"* Behaviour & Information Technology (BIT), *vol. 41, no. 10, pp. 2137–2159, 2022*. DOI: https://doi.org/10.1080/0144929X.2022.2068070

Supplementary Information The online version contains supplementary material available at https://doi.org/10.1007/978-3-658-47445-4_8.

S. Linsner, *Privacy Preserving Data Management*, Technology, Peace and Security I Technologie, Frieden und Sicherheit,
https://doi.org/10.1007/978-3-658-47445-4_8

8.1 Introduction

Collaboration and digitized work processes require the collection, processing, and exchange of data [104]. However, not all data should become available for everyone. The value of privacy protection often depends on the circumstances of use, as Steinbrink, Reichert, Mende, and Reuter [105] has shown by the example of the behavior of refugees during flights and its impact on the management of personal data. In business contexts, it is necessary to administrate the access to operational data in order to provide third parties with necessary information, but without becoming too transparent for competitors [106]. Negative consequences of data disclosure in agriculture can be very severe. Gupta, Abdelsalam, Khorsandroo, and Mittal [107] summarize some consequences of data leakages as potential threats to modern farming ecosystems, like competitors or hostile actors taking advantage of that information. Farms usually are small and medium-sized enterprises (SMEs) according to the definition of the European Union (EU)[1]. Many farms are even entirely family-run [108]. The weaker market position of these businesses is not only noticeable in the supply chain, but is also reflected in the technology development: Innovation is driven by the big players in the supply chain, often leaving perspective and needs of SMEs, e.g., in agriculture, underrepresented [1, 109]. This especially holds true with regard to IT solutions. There are many tools available for farmers to manage their operations [1, 110]. However, these mostly require the farmers to upload their business-related data to third party servers. In doing so, they often permit the platform-providing companies to use and sell their data by accepting the terms of use [109]. Becoming too transparent to competitors or other actors along the supply chain is problematic for farmers: If their operations are visible to others, they might get bought out by bigger companies that use the operational data of the former landowners and cheaper workforce to cultivate the land for a greater profit [20]. In order to prevent the consequences of unwanted or excessive data disclosure, transparency could be beneficial for end users. By making the disclosure process more transparent and highlighting the data that is made available for others, mistakes and unintended data leakage could be prevented. However, increased transparency could also lead to information overload for the operators. Especially in agriculture, the owners of small farms have tight time schedules to fulfill their work routine. Office work is not the core of that routine and has to be done efficiently, often in the evening. To assist users in making informed decisions about the disclosure of their

[1] https://ec.europa.eu/growth/smes/sme-definition_en, based on headcount (micro < 10, small < 50, medium < 250) and turnover (micro $\leq$ 2 million €, small $\leq$ 10 million €, medium $\leq$ 50 million €) of the enterprises.

data, we integrate and evaluate transparent controls to raise awareness about data sharing scenarios.

In this paper, we investigate the current status quo of data sharing practices in SMEs in agriculture by conducting a qualitative prestudy including 52 German farmers. Based on this, we derive requirements for confidentiality preserving data sharing in agriculture. To investigate whether transparency eases the task of data disclosure for business cooperation or is a mere data overhead that leads to information overload, we evaluate our system with nine users from the agricultural domain and, in contrast, nine users without agricultural experience. Due to the comparable nature of the disclosure of sensitive business data by SMEs and the disclosure of personal data of end users, the results could also be applicable in privacy-related contexts.

The structure of this paper is as follows: Section 8.2 gives an overview of related work and background information and identifies a research gap. In Sect. 8.3, we derive our research approach and formulate two research questions. Subsequently, we present our qualitative study with 52 German farmers and investigate data sharing scenarios in Sect. 8.4. In Sect. 8.5, we derive requirements for information systems in agriculture based on the prestudy and present a concept to address these. The evaluation of the system will be presented in Sect. 8.6. Section 8.7 discusses our results and concludes this paper.

8.2 Related Work

In this section, we present related work on two main topics: Firstly, we present current research on the state of digitalization and data sharing in agriculture. Together with the results of the prestudy presented below (Sect. 8.4) the literature is later used for the identification of requirements (Sect. 8.5.1). Secondly, we elaborate on advances in the HCI community on transparent user interfaces, which could apply to our use case of data sharing for non-expert users. In this way, these works contribute to the design of our data sharing concept (Sect. 8.5.2).

8.2.1 Data Sharing in Agriculture

Regarding the question of how farmers perceive the processes of digitalization in their domain, several studies exist: Fountas, Blackmore, Ess, Hawkins, Blumhoff, Lowenberg-Deboer, and Sorensen [68] asked 198 farmers in the U.S. and Denmark about their attitude towards precision agriculture. They found that main obstacles existed in too time-consuming data handling and that 80% of farmers wanted to store their data locally on self-hosted servers. Carbonell [69] identifies power asymmetries between farmers and big agribusinesses as a problem. The author sees open source tools and open data as promising approaches. An overview of the adoption of digital tools by 287 participants from agriculture in seven EU countries is provided by the study of Kernecker, Knierim, Wurbs, Kraus, and Borges [70]. It revealed that farmers need better instructions and security. It also became clear that most farmers with more than 500 ha land run fully digitalized businesses, whereas smaller farms still lack digitalization. The fact that in particular small enterprises lack digitalization has also been concluded by Annosi, Brunetta, Monti, and Nati [111] and Regan, Green, Maher, *et al.* [71]. Regan, Green, Maher, *et al.* [71] illustrate this by means of agriculture in Ireland, which consists largely of family-run farms. The authors present an interesting view on data ownership and privacy of farmers. They found a general distrust towards companies, but a very open attitude towards actors with whom the farmers had longstanding partnerships. The authors assume that the reason for this is the family-owned business model. Furthermore, previous work has outlined that access to corporate data is an existential problem for farmers, as noted by Fraser [20]: Increasing "data grab" can lead to "land grab". Once companies have access to business data, they can easily take over the farm by estimating the effort and profit of the farmers and buy them out. By acquiring many smaller farms, companies can manage large-scale agricultural businesses with the data they obtained from former owners. With less effort, the companies are able to gain much more profit from the land than many small farms before. In this context, Ferris [52] sees opportunities in technologies for precision agriculture, but also discovers dangers that arise from massive collection of data: Exposure of personal data, income, or yield of the fields. Ferris states that farmers fear disadvantages if this data is accessible to their competitors. Therefore, the author calls for governmental regulation. The fact that this affects not only individuals but an entire domain is supported by the demographic situation in Germany, as shown by federal statistics [18]: The majority of farmers cultivate less than 200 ha of land. This underlines the need for research on SMEs in this domain and their role in business collaboration and data exchange. Utilizing transparency for user interfaces is a possible solution, as shown in the next subsection.

8.2.2 Assisting Users with Transparency in Data Sharing Scenarios

For many users of data sharing applications, the concept of individual privacy and corporate data sovereignty often remains abstract. At the same time, the process of viewing data or the act of granting access rights are very concrete issues, which have monetary consequences—particularly in business-to-business (B2B) constellations. Visualization as a means to better communication and incentive for conscious data sharing decisions can lead to an increase in general trust towards data recipients and prevent cognitive overload among users [112–114]. Nevertheless, it was found that the extension of choices in some cases leads to an unintended, opposite effect, called the control paradox [34, 115]. In these constellations, users who have more confidence in controlling their data tend to disclose more than they would have without transparency supporting technology. Therefore, improving transparency does not necessarily lead to advantageous decision-making by users. Mitigation of the control paradox should be considered when designing transparent tools. Guidance for this process can be found in many privacy patterns from existing literature. An overview of existing patterns can be found at Iacono, Smith, and Gorski [116]. Patterns which are relevant for our implementation are referenced in Sect. 8.5.2.

A rich body of literature exists on the visualization of privacy aspects in different contexts (e.g., social media [117–121]), but authors rarely refer to other fields of data sharing. Overall, research on the visualization of privacy and corporate data sovereignty seems to be intertwined with the presuppositions, methods, and theories of several disciplines. This has therefore led to very different approaches in the visualization of data sharing between companies. In the following, we aim to organize the diverse literature to identify the research gap that our paper intends to address.

In general, the complexity of collaborative work leads to less overview and clarity than in the classic provider-user relationship, as found, e.g., in the selection of cookie policies on websites. Moreover, within supply chains, the trade-off between closer networking in collaborative systems and security and privacy aspects is more substantial, as already described by Tolone et al. [122]. They found that collaborative systems are generally characterized by a higher quantity of users and operations, as well as more variation in enterprise models, object types, and tasks. Compared to single-user systems, these features make it more challenging and time-consuming to implement discrete transmission paths. In their comprehensive review on future security and privacy challenges in agriculture, Gupta et al. [107] note that the complexity of collaborative data sharing will further increase with the adoption of smart farming hard- and software. Considering the collaboration between lay and non-expert users, which is a common characteristic of many SMEs, the dire need for an understandable and transparent presentation of complex data sharing constellations becomes apparent [123]. In the following, we will present those approaches from research that can be classified as privacy-enhancing technologies. Please note that technologies which help to provide or enhance privacy for personal data may also be used to protect sensitive data and prevent involuntary disclosure in business contexts, even though business data is not subject of privacy per definition. Also, the presented collection of approaches does not claim to be exhaustive in this regard, since, especially in the B2B context, commercial platforms are often used for data sharing, to which access is only possible via subscriptions. Rather, the compilation is the result of a thorough analysis of existing academic literature, where we cluster the different approaches for data sharing policy visualization into three categories: access control matrices, network approaches, and privacy dashboards (see examples of each category in Fig. 8.1).

First, the access control matrix represents one of the initial forms of access rights management [123–125]. It provides potential users with access rights by assigning the access to certain sensitive objects (y-axis) to specific actors/users (x-axis) in the form of a simple table (see Fig. 8.1a). From this starting point, the concept of the access control matrix has been further developed by different approaches. Using shared data in social networks as an example, Hu et al. [126] developed a color-coded representation of conflicting privacy settings between linked users. This proposal was complemented by two pressure indicators as metaphors in the tools' interface, which allows users to weigh the privacy risks against the loss of audience reach. Through the application "Expandable Grid", Reeder et al. [127] also propose a modified access control matrix in which colors stand for different, potentially overlapping sets of data access rules. Conversely, the proposal by Kolomeets et al. [128], which was developed for hierarchical contexts within a company, resorts to triangular matrices. Here, different colors were assigned to different categories of access rights, adding the distinction between reading and writing rights to the binary concept of access versus denial [128].

Second, network diagrams form another large cluster of proposals for visualizations of data flows (see Fig. 8.1b). The nodes usually represent the entities which have data in- or outflows. The edges represent the data paths (see e.g., [129, 131]). Depending on the approach, edges are modeled as directed or undirected. By assuming directed edges, it is implied that data is sent only one way, and data exchange cannot take place in both directions. For example, in Wang's proposal for a hierarchical access role system, directed edges are adopted [132]. For scenarios of collaborative work, instead, it is more appropriate to apply undirected edges, since transparency is created here by disclosure of certain data between entities in the

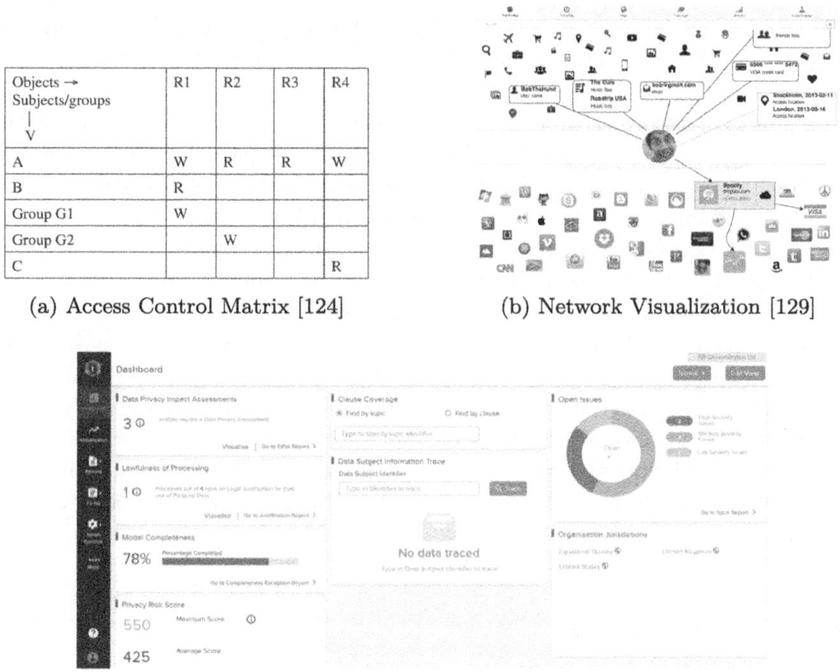

Objects → Subjects/groups ↓	R1	R2	R3	R4
A	W	R	R	W
B	R			
Group G1	W			
Group G2		W		
C				R

(a) Access Control Matrix [124]

(b) Network Visualization [129]

(c) Privacy Dashboard [130]

Fig. 8.1 Examples for the three clusters of data sharing policy visualization

process of mutual data exchange. An example of representation via an undirected network is given by Chou et al. who model a social network [133].

The third and final cluster comprises privacy dashboards, which generally tend to focus on overall assessments of data exposure risks rather than on the depiction of data flows themselves (see Fig. 8.1c). Examples of this approach include the work by Wilkinson [134] and Bagnato [135]. It is conceivable that dashboards could include aspects of both of the aforementioned categories. The "Impromptu" prototype by Rode et al. [136] visualizes privacy via a map of concentric circles on which files can be interactively mapped, and multiple levels of warning about unwanted data exposure have been integrated. Current projects such as "TrUSD" [137] and "Poseidon" [135] show that there is currently a trend towards visualizing data transparency more clearly by using dashboards, although research in this area is still in its initial stages. As Tolsdorf et al. [138] pointed out recently, implementations of privacy dashboards do not yet comply with the General Data Protection Regulation (GDPR) by the European Union, leading to information deficiencies for lay users.

While applying different methods of evaluation (see 8.6.1), all of the studies on the various visualization approaches reported positive impacts on the data sharing decision-making. Differences also exist in the various formats of visualization concerning the point in time at which they become visible to the user. Some proposals work with dynamic live visualizations, some of which are interactive [129, 136]. Other approaches offer post-decision visualization as feedback, such as the tool by Vaniea et al. [139], where the output of privacy policy decisions and arising conflicts are presented in tabular form. However, their tool explicitly targets professional IT expert groups, which contradicts the focus of this paper. Meanwhile, the analysis by Wilkinson et al. [134] is explicitly aimed at non-expert users and tries to achieve an improved understanding of data exposure through an interface that is as transparent as possible for current data extraction. Kolter et al. [140] suggest a different approach by presenting solely the visualization of data streams that took place in the past user history.

8.2.3 Research Gap

In agriculture, the exchange of data is an important factor in state-of-the-art work routines. Digitized machinery operates on the basis of pre-processed data [110]. Moreover, division of work is a widespread practice due to the fact that most farms in Germany are SMEs or even family businesses and only able to gain access to expensive, new machinery through cooperative business models [48, 108]. Accordingly, technological innovation is mainly driven by large companies which offer their products to farmers [37]. For using the products (e.g. crop management or telemetry tools), farmers are often required to upload their operational data to third party servers [20]. Subsequently, they no longer have control over their data. There are also services that offer free tools to attract farmers who otherwise could not afford expensive enterprise products. However, by using these services, the farmers grant major upstream companies access to their farm data and plans for the current season [1]. In order to identify important aspects of farmers' perspectives on data sharing in their domain, a study with affected farmers, especially from small farms, is important.

Transparent user interfaces for farmers could be a promising approach to grant individual farmers more control over their data. The positive evaluations of the diverse visualization approaches mentioned in 8.2.2 showed that transparent controls in data sharing applications could help to prevent unintended data leakage [127, 128, 132, 136]. Data loss in agriculture is often caused by disadvantageous interfaces or centralized cloud solutions where farmers have to upload all data to third party servers, which may even be affected by non-EU laws [141]. However, transparent user intervention poses the risk of information overload. Related work tested transparent software technologies with user samples consisting mostly of university students in tech majors, e.g. [127, 129, 132], despite the fact that Parker and Sinclair [142] identified the importance of actual end user-centered design in agricultural IT as early as 2001. Farm managers and employees usually do not have expert computer skills, and their workflows often suffer from tight time constraints [1]. Since many of the evaluations were conducted with IT expert users [128, 132, 140, 143], we can hardly apply their results to our target group.

Another aspect concerns the fact that most of the literature cited in this section deals with visualizations of privacy and transparency in the private sphere. Despite an extensive literature search, no work could be found that explicitly considers the matter of usable visualization in the B2B context of supply chain dependent SMEs. Consequently, a research gap in the use case of non-IT-expert users in information sensitive collaborative commercial contexts can be identified. Therefore, investigations on the effect of transparent user interventions on users from SMEs such as farms are urgently needed.

8.3 Research Approach and Questions

To address the lack of lay user-specific, transparent data sharing solutions for SMEs identified in related work, we pursue to answer the following research questions:

1. [**RQ1**] Which requirements exist for agricultural information systems to allow small enterprises to participate in cooperative work scenarios without becoming transparent for competitors?
2. [**RQ2**] (How) Can transparent controls contribute to better privacy behavior for farmers who are often time-constrained due to their daily work routine?

The research approach of this paper is illustrated in Fig. 8.2. In this work, we present an approach consisting of several steps: After our literature review presented above (Sect. 8.2), we share the results of our qualitative study with German farmers to shed light on data sharing practices from their work routine (Sect. 8.4). Here, the perspective of SMEs is highly relevant, as, in contrast to large enterprises that can afford dedicated departments for different activities, e.g., a legal department, small enterprises rely on workers who usually have to perform many different activities. This heavy workload reduces the time available for carrying out office tasks and thinking about the consequences of possible data flows. For this reason, awareness of data protection in the exchange of business details probably differs between SMEs and large enterprises. Therefore, SMEs especially need easy-to-use and easy-to-understand solutions that set the bar high for the development of an appropriate user interface. Furthermore, since the conflict between the time one can spend in the office and the time available for practical activities is particularly challenging in agriculture [68, 144], our assessment focuses on employees in small farms. However, the results could also potentially apply to other sectors.

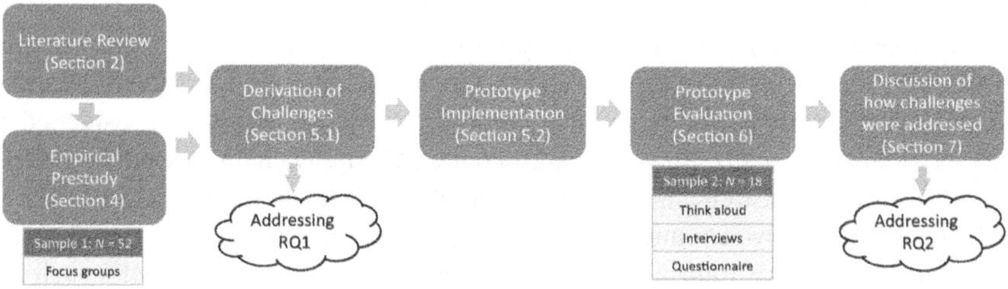

Fig. 8.2 Flow chart of the research approach

Second, the results of the prestudy are used to derive challenges and requirements regarding data sharing applications and a data sharing model designed to address these challenges (Sect. 8.5), intended to answer RQ1. We then use the identified model properties to design and implement a interface prototype which enables data sharing within an agricultural context.

In a third step, we evaluate the prototype qualitatively to investigate user behavior in data disclosure scenarios and the effectiveness of transparent control options (Sect. 8.6). Last but not least, we will discuss the results of the prototype evaluation with regard to the previously identified challenges, thereby answering RQ2 (Sect. 8.7).

8.4 Empirical Prestudy

This section presents the prestudy we conducted with members of the target user group to gain insight into data sharing in agricultural B2B relations, to enable us to derive requirements for our data visualization model, and to give a general overview of digitalization in agriculture.

8.4.1 Method

This subsection covers the methods applied for our prestudy. The entire process, comprising creation of an interview guideline, recruitment, conduction of the focus groups, and data analysis and storage, followed the guidelines of the ethics committee of *Technical University of Darmstadt*. Previous to this contribution, we have published a paper on the current state of digitalization in agriculture, based on the same data [1]. One of the various concerns about digitalization in agriculture was data security. Frequent mention of a lack of tools for sharing data for decision making motivated us to elaborate on the specific needs of data exchange between farmers and develop our own solution approach. In this paper, while analyzing the same data in the empirical prestudy, we focus on the future and on demands for upcoming approaches, as well as topics relevant for HCI. We specify the observed subtopic on the demand for usable applications in order to design an implementation application (Sect. 8.5.2), which is subsequently evaluated (Sect. 8.6). These contributions have not been published before. Therefore this paper differs substantially from the previous work and builds on it towards designing mechanisms for usable security.

8.4.1.1 Study Design

We identified focus group interviews as the adequate method for the prestudy. This method consists of analyzing guided expert discussions in groups. Focus group interviews [30, 31] belong to the standard of qualitative-exploratory research approaches. The group approach enables a conversation between experts in which the researcher adopts only the role of observer and interested listener. To support the participants in case of a halting conversation, we created a set of supporting impulse questions, which only had to be resorted to in a few instances. Readers can find the interview guideline in the Electronic Supplementary Material A.1.1. It only includes the guiding questions relevant for this publication. Due to interview economics we also discussed resilience capacities of the farms [11] and blockchain technology in their domain. For the latter, a brief explanation was offered. To keep the subjective bias as small as possible, the focus group interviews were guided by two researchers experienced in conducting interviews and the focus group method. As the interviewed farmers are located at a variety of places, our project partners identified occasions when several potential participants would gather to conduct the interviews. Finally, we ended up with 12 focus groups: Nine at an education center for aspiring agricultural technicians (FG1-FG9), two at an innovation center of a farming machine manufacturer (FG10 and FG11), and one at a machinery ring (FG12). We conceptualized all phases of the study based on the guidelines for ethical research of the ethics committee of *Technical University of Darmstadt*. These guidelines involve a self-assessment checklist of the risks for the participants that turned out unproblematic. Therefore, no further actions were required to to satisfy the ethics committee's requirements.

8.4.1.2 Participants

The sample consisted of N = 52 farmers. All surveyed farmers hold responsibility and decision-making power in SMEs, as family farms dominate the land structure in the studied region. With 45 people identifying as male and seven identifying as female, a balanced gender distribution could not be achieved, but the proportion corresponds to the share of female managers in the agricultural industry [108]. Due to predominantly younger participants at the educational center (42 in the segment 20-30), the age average of our study is significantly lower than the average age of the agricultural labor force. For an overview of the fields of work of our participants, see Table 8.1. Participation in the focus groups was voluntary and unpaid. We ruled out overlapping of the prestudy sample with participants in the later described usability study to provide an unbiased usability analysis.

Table 8.1 Branches the participants work in (multiple possible)

Branch			Amount
Cultivation of grain			22
Viticulture			3
Cultivation of vegetables			1
Husbandry	Beef raising	Dairy cattle	12
		Breeding	4
	Pig housing		4
	Laying hens		3
	Biogas production		3
Service provider			6

8.4.1.3 Data Analysis

The focus groups were conducted in German as this is the native language of all participants. Audio recordings were made, which we then transcribed and anonymized. The length of the group sessions varied between 15 and 27 minutes. Using the open coding method of grounded theory, we formed text segments [81], which were then categorically coded to assign them to specific semantic categories. The classes could be expanded during the process, whereby all previously coded segments were re-examined. In line with the quality criterion of intercoder reliability, two of the authors conducted the classification in constant consultation to minimize subjective bias in the coding. Ultimately, the categories and assignments were cross-checked by the team of authors. The English translations of segments cited in the results section are as literal as possible. Participants often used names of companies (machinery manufacturers, suppliers of feed, fertilizer and seed, contractors, laboratories). These were replaced by general descriptions that do not allow the readers to draw conclusions about a specific company.

8.4.1.4 Limitations

Limitations in the prestudy are mainly rooted in the sampling. Most importantly, we cannot rule out a gender or age bias. Additionally, the regional distribution of the experts is skewed towards the former Western German States. For an in-depth discussion of these limitations, we refer the readers to Sect. 8.7.

8.4.2 Results

To ensure an improved data model, this result section initially presents our findings and reveals the challenges farmers face in the present state of agriculture. The section starts with a general description of the digitalization status at the whole sector level by briefly summarizing the results of the previous contribution [1], to aid the reader in understanding the topic. As a second step, we derive new and original insights based on the material on specific characteristics of data sharing in the sector (subsection 8.4.2.2). Moreover, new challenges are identified that currently hamper data sharing for farmers (subsection 8.4.2.3).

8.4.2.1 Agricultural Digitalization

First, we turn to the status of digitalization in Germany's agricultural sector. The participants' responses paint a differentiated picture of the advantages and disadvantages of digitalization. The level of experience with digital tools among the participants ranges from occasional data entries on stationary PCs to advanced on-farm network architectures. In addition to home PCs, most participants primarily use mobile devices such as smartphones, tablets, and embedded computers on large-scale equipment. Windows 10 and Android are the most used operating systems, but Windows 7 and macOS/iOS were also mentioned in a few cases. All focus groups identified certain advantages and disadvantages, which are listed in the following. Databases and accounting programs offer assistance in this respect and make the formerly handwritten work of farm management and bookkeeping easier and more comprehensible. In addition, the physical work on the farm changes as a result of digitalization, as it is an essential prerequisite for smart and precision farming. Both farming strategies promise efficient and reduced resource use, improved yields, and environmental friendliness in the process. Or concisely in the words of a participant:

> Q1 (FG2): *Be more productive, save resources, and be more environmentally friendly.*

We learned that precision farming uses accurate global positioning, area-wide sensor networks (weather, soil), and automated equipment (vehicles, fertilizer distribution, irrigation) to achieve the highest possible efficiency in the cultivation process. The data generated by these processes is hereafter referred to as operational data. For example, by relying on geodata stored in digital orders, tractors already drive with automatic steering at many of the respondents' farms. For some operators, time-consuming and monotonous work at the steering wheel is thus transformed into monitoring while simultaneously completing administrative tasks.

> (FG5): *Driving with automated steering systems is, of course, truly advantageous and practical, saves time, money, and protects the environment.*

In contrast, other participants argue that increasing automation can change mentality if too much trust in and reliance on precision farming leads to unintended inattentiveness.

> (FG11): *Automation means that one is no longer as focused on your work as one used to be. You drive down the field, on the way you make three phone calls, write another mail, because the technology does it all automatically - That's one problem with automation, that you then rely too much on the system.*

Overall, the focus groups agree that the job profile of a farm manager will undergo major changes in the future. According to them, there will be greater demand for training on technical equipment, and the classic image of a profession in which workers spend relatively little time on administrative work will be blurred. According to several focus groups, there is also a tendency for the fairly small-scale farm structure in the examined region to give way to larger cooperatives or individual companies.

> (FG4): *If the process continues like this, it will lead to the small ones quitting the market more and more, and the bigger ones will need to merge or cooperate.*

In the interviewees view, this conceivable structural sector transformation lacks behind, on the one hand due to the average age of farmers on small farms and their unwillingness to integrate new technologies into operational processes. This opinion of our significantly younger sample of interviewed farmers could be subject to age bias. On the other hand, participants mentioned high investment costs, which usually pay off the sooner the larger the farm revenue is.

> (FG7): *The small sideline farm with 20, 30 acres will not purchase a smart farming tool because it does not pay off financially and has no significant benefit. But on large farms, smart farming is already part of everyday business.*

Then again, the scenario was described that in the long term, smart farming technologies would also become more affordable for farms with less than 50 acres of arable land due to rising proliferation and lower production costs. Furthermore, part-time farmers are reluctant to digitalize their farms. Most focus groups agreed that digitalization is mainly a competitive burden for smaller farms – and a worthwhile investment for large farms or cooperatives. Digitalization consequently provides some disadvantages in addition to the advantages as mentioned earlier, which the focus groups identified through differentiated discussion.

8.4.2.2 Data Sharing in the Agricultural Sector

As previously noted, agriculture became a data-intensive sector due to its collaborative structure and complex supply chains (8.2.1). For more insight into current data sharing behavior, focus groups discussed the media used and expressed needs and critical issues. The majority of farmers communicate with business partners through online media such as email and messenger services. The latter was highlighted several times as being particularly practical, as it offers the possibility of easily sending images and locations. Depending on the urgency of a matter or the proximity of the business partner, respondents also find face-to-face meetings and phone calls convenient, especially if they have built up a certain level of trust through decades of collaboration.

> (FG8): *I am a technical advisor for a large seed community, and we actually still do business personally. Also by phone, of course, but you need trust in agriculture, which is not easy to build up.*

Field data is frequently transferred via USB stick to be compatible with the automated equipment. Print letters and faxes are a rare exception, mentioned at best in the context of contracts. For extensive transfer of field data, the latter two options are barely chosen anymore.

There is an awareness among all participants about the intrinsic value of farm data and its potential utility for companies upstream and downstream of the supply chain and the associated risks.

> (FG9): *I believe that many companies are interested in the data. I have seen offers that a digital crop field card gets cheaper by ten euros a month if company XY is allowed to look into the data.*

In most focus groups, skepticism about the intentions of other companies about what happens to the data after it is transferred and whether it is sold to third parties, predominates. Profiles derived from the data could result in individualized prices, which in turn would lead to reduced profits. Some focus groups took the line of argumentation a step further and warned of the risk that companies could draw numerous conclusions from comprehensive and multi-year farm data that experience-based workflows could be inferred from it. In particular, for smaller family-run farms, experience in farming and knowledge of the land is considered one of the most critical assets. The more data large companies have pooled, the more likely farms are to be vulnerable to takeovers.

> (FG5): *Now, many companies want to have our information on how we proceed in the field or what we have learned over the years. [...] Through the many years of experience that we have built up, if someone has the same information, we could then be replaced.*

Two focus groups discussed the financial value of the operational data. According to them, current compensations from chemical companies and equipment manufacturers tend to provide access to data platforms or fertilizer recommendations at a lower price. Some participants see data as another profitable resource with its own supply and demand dynamics. Risks from disclosing operations should be compensation by fair quid pro quos:

> (FG9): *As soon as other companies want to have the operational data, then they can, of course, also get it at a certain price. Depending on what kind of data it is, there has to be corresponding compensation.*

In-house, some farmers face data overload due to a lack of compatibility between different systems and platforms. Data sets would need to be prepared differently for automated vehicles, regulatory applications, or contracting orders, creating significant additional work for farmers. For example, one participant complains,

> (FG11): *that we produce an incredible amount of data junk and that us farmers have between two and five different systems which we have to maintain separately.*

On the contrary, one of the participants envisions an extended mode of data sharing on a cooperative level between farms for mutual benefit. Data overflow or data demand could depend on the size of the farm and its digitalization status. However, our focus group method did not allow for an examination of this connection in individual cases. Not only does the quantity of data burden some participants, but the quality of the data collected by automation may also be a concern. German authorities closely monitor the application of certain fertilizers and pest control. Forwarding data that an automated device may have incorrectly collected can result in the inference of a violation of regulations. Checking the data, however, takes a similar amount of effort and time as the conventional manual entries:

> (FG11): *We don't save time; we just distribute it differently. First of all, we have complied with the law by documenting fertilization directly, but afterward, we have to check whether what we have documented is also what we have done outside.*

According to the focus groups, increasing carelessness while operating automated vehicles (see quote 2) is part of the problem. Time savings and trust in automatically gathered data and its direct forwarding to authorities are consequently rated low in the focus groups. Involuntary data disclosure through hacking attacks—especially ransomware—was also identified as a risk, increasing alongside digitalization.

8.4.2.3 Current Challenges in Data Sharing

We identified five areas that represent challenges for data sharing for the model from the focus groups. First, many participants are bothered by the fact that a lack of compatibility of different systems limits the current state of farm digitalization. One reason for this is that large manufacturers are trying to lock customers into individual product families.

> (FG1): *But once you have started with [machinery supplier], you are also dependent on their system. [...] With the high follow-up costs, there is also the question of whether you can ultimately increase yields or quality.*

Second, the focus groups reveal how diverse the agricultural sector is. Different types of farms, farm sizes, crops, and levels of digitalization face different specific data sharing risks and data ownership expectations. Some farmers are in a dual role, acting as labor or equipment rentals, consultancies, or vendors in addition to their production. Third, farm management and employees use very diverse forms of media to communicate and share important data. Fourth, data sovereignty over critical agricultural data is threatened by rising data demand from both upstream (machinery producers, biochemical companies, and seed producers) as well as downstream entities (retail, consumers) of the food supply chain. While specific data sharing is necessary for the fulfillment of contracts, complete transparency of farm processes is perceived as weakening the market position, as the risk for SME farms to be bought up by larger companies increases. Lastly, to be able to provide evidence in legal proceedings following data breaches, subsequent traceability of sharing activities is perceived as helpful. Current practices, involving mail or cloud solutions, only allow for this to a limited extent.

8.5 Designing Transparent Controls for Data Disclosure

This section describes the challenges derived from the prestudy and the concept as well as an implementation of a prototype for a formalized data sharing control. First, it is explained how the problems identified in the prestudy (see Sect. 8.4.2) can be addressed and which requirements should be included in the solution concept. Since farmers demand more control over their own data, a prototype is implemented to investigate whether offering transparent control options actually helps users in this regard. Since data management by centralized third party applications has been criticized, and measures to protect data ownership have been demanded, an independent and decentralized system is targeted in our concept.

8.5.1 Challenges, Requirements and Concept

Based on the prestudy (Sect. 8.4.2) and relevant literature (Sect. 8.2.1), we were able to derive the challenges of (1) Compatibility, (2) Sector Diversity, (3) Media Diversity, (4) Transparency Balance, and (5) Retraceability. We then derive corresponding requirements for data sharing platforms and data disclosure scenarios, as well as a respective concept addressing them. These are listed in Table 8.2.

The *first challenge* arises from the lacking compatibility of existing products, which leads to vendor lock-in effects. To address this problem, the use of open-source tools and standards is recommended—a claim which can also be found in related work [69]. Furthermore, the system should be independent of a specific im-

Table 8.2 Challenges and requirements for data sharing platforms

Challenges	Literature and Prestudy	Platform Requirement
Challenge 1: Compatibility	Soft- and hardware of different services and product families are not compatible because of differing data formats (Q10) and lock-in effects for machinery (Q12) [1, 69].	Application of standard data formats to facilitate compatibility across products. Independence of internet connectivity through local accessibilty.
Challenge 2: Sector Diversity	The agricultural sector and the food supply chain is characterized by a multitude of actors and roles (Q5, Q6), [69, 107].	Flexibility of the platform to account for different actors and roles.
Challenge 3: Media Diversity	Farmers have a broad spectrum of media usage, ranging from pen-and-paper, telephone and fax to PC and mobile phone (Q3, Q4, Q6). The level of digitalization often depends on the size and type of the farm [1, 70].	Support of interfaces on desktop PCs and mobile phones as most commonly used media.
Challenge 4: Transparency Balance	While greatly demanded from upstream and downstream actors, full transparency of farm processes may lead to SMEs finding themselves in disadvantegous market positions towards larger companies (Q7-Q9) [52, 70, 71].	Improvement of data sharing choices and provision of job specific approvals. Restrictive pre-selection to meet privacy by default.
Challenge 5: Retraceability	Currently, not all data disclosures can be traced in retrospect, which is a hindrance to the legal sanctioning of data breaches by the receiving parties (Q11) [52, 107].	Ex-post traceability of data sharing decisions and contents provides evidence of all transactions for legal proceedings.

plementation or background technology and enable local processing of data sharing decisions. This way, the model can be applied to various technologies: Data exchange via email is supported as well as client-server architectures, cloud-systems, and blockchain ecosystems. The latter are experiencing a boom in the industry and are seen as a promising approach for the food industry, as they offer transparency along the supply chain and not merely simplify quality assurance, but can also prove the origin of products to end consumers. Therefore, when designing systems, they should be compatible with blockchain architectures.

The *second challenge* concerns sector diversity, since actors and roles in agriculture are highly diverse and can change over time. This requires individual data disclosure for each actor in changing situations. To meet this requirement, form-based workflows could be useful. Templates could be provided for different workflows and complemented by different actors. If access control matrices are used to customize data disclosure, the addition of further parties in a contract would be easy. For each participant, a new row has to be added. For further rights of use, a new column has to be added.

Addressing the *third challenge* of media diversity, this requirement must be extended in regard to the problem of sector diversity. Because of the huge differences in procedures and media used, workflows within a data sharing platform should be adjustable on the fly, for example while driving a machine in the fields. Regarding the media, farmers indicated PCs and mobile devices as the most common media. Support of both media types should be available as a minimum requirement for a successful platform implementation. Responsive tools should be used to provide workflows that are usable on every device.

As for the *fourth challenge* of transparency balance, data ownership was identified in the prestudy as one of the core problems in current data sharing scenarios. For farmers, who often rely on third-party services for planning, management, and task execution, data sharing is indispensable. This gives third parties access to planning and accounting data. In an independent system, users should be able to decide for themselves who gets access to which data. Data owners should be empowered to control who receives data, and purposes for using the data should be documented. This confirms the findings of Fountas, Blackmore, Ess, Hawkins, Blumhoff, Lowenberg-Deboer, and Sorensen [68] that farmers prefer local storing and processing of data. To make the control over the data flow more transparent for users, form-based workflows with built-in transparent control features should be used for data disclosure. To minimize the number of actions the user has to fulfill, we propose that the items in a form are sorted and grouped in semantic units. Each of these units is given a control option to configure which participants in the order are allowed to see the data and which are not. We describe a concrete example implementation in more detail in Sect. 8.5.2.

The *fifth challenge* deals with ex-post traceability of data. Since data cannot be retrieved once it is made public, compliance is very important. Lawsuits may arise if data is misused, which requires permanent and tamper-proof documentation of granted rights of use. These rights have to be defined in the moment of data disclosure and stored in a tamper-proof manner. To provide evidence in court in case of data breaches, the final requirement is the ex-post traceability of shared data by documented rights of use, whose integrity is proofed by the system. If a blockchain architecture is used, these can be documented in a tamper-proof manner. However, this is in conflict with applicable deletion periods. To solve this problem, various technical precautions should be taken: First, it must be possible to delete the data in a timely manner. For this purpose, reliable maintenance mechanisms must be in place for client-server architectures or cloud systems. In the case of a blockchain environment, this is more difficult, as integrity is an essential aspect of the blockchain. However, since this paper presents an approach that is neutral to specific implementations, this problem exceeds the scope.

8.5.2 Implementation

We implement and evaluate a prototype to gain empirical evidence on the willingness to define usage rights and on the perception of control over the disseminated data, and to explore how the previously identified challenges could be addressed. A conceptualized view of the data processing can be found in Fig. 8.3. The prototype consists of two parts: First, there is a form service (*"To complete"*), which enables the user to fill out provided commission forms for business partners and to define rights of use (Fig. 8.3, *Define*). Based on the inputs of the users, a splitting script creates separate views for each participant in the form of JSON documents (Fig. 8.3, *Process*). In this way, it is ensured that only the designated data leaves the local storage. The second service for viewing (*"Check"*) can use the exported files to render an individualized form for each recipient based on the selected data (Fig. 8.3, *View*). As mentioned in the previous section on the fifth challenge, the third step within the data flow (Fig. 8.3, *Transfer*) was out of scope for the prototype.

The prototype was implemented as follows: Both services were embedded into a single progressive web application based on `Ionic/Angular`. In this way, the app could be delivered as a web page for remote evaluation purposes, but also later embedded in a standalone app for mobile devices or desktop computers, which addresses the third challenge of media diversity. The form rendering was done with

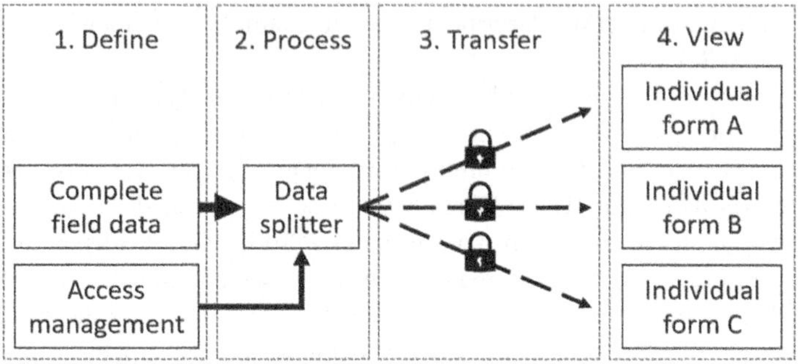

Fig. 8.3 Derived concept for required data flow

the npm package `formio/angular` of the *form.io* project. Via a floating menu at the bottom of the screen, the user can switch between the two services. For the evaluation, the application was hosted within a docker container on a server. However, a local instance would be viable in a real-world application. The form service requires the user to load a JSON template for a form and provides the possibility to load another optional JSON file to load previously selected values. This way, the farmers would have the possibility to reuse the planning from the year before for convenience reasons. The loaded JSON files would then be used to render the form, which could then be filled out by the user. Using a template approach, we took up the second challenge to enable sector diversity. The then rendered form provides access control matrices to define the visibility of data for each data category (e.g., details on seeding material, required equipment), addressing the fourth challenge of transparency and the privacy pattern of "Direct Access to UI Components" [116, 145]. Users are presented with the control options close to each data item to ensure that they can immediately specify who could get access to that data. Therefore, the control matrices show the different involved parties for the job in rows with radio buttons to select the data visibility for each party. Examples are shown in Fig. 8.4.

If no previously selected data was loaded, the default value for data visibility is *"must not see data"*. In this way, an approach was chosen that preserves privacy by default, supporting the privacy pattern of "Disable by Default" [116, 146]. After completing the form, the user can press a button *"Submit"* to process the data. When clicked, the data is processed in a splitting script, generating two JSON files for each recipient. One contains the disclosed information and the second a corresponding form to display the values properly. Within the prototype, these files could be exported and distributed (e.g. via email, blockchain, clouds, client-

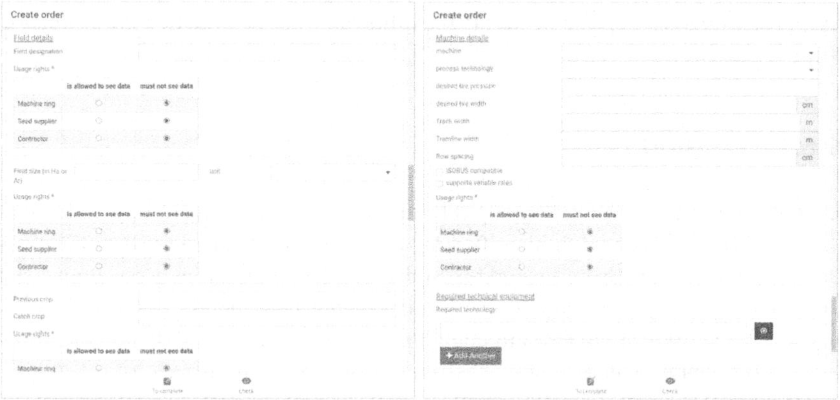

(a) Section about field details (b) Section about machine details

Fig. 8.4 Screenshots of the prototype with examples for control matrices in one loaded form

server). By using a standardized format such as JSON, we wanted to address the first challenge of compatibility, since it enables an easy implementation or conversion to another data format.

The exported files could now also be used to be loaded into the check service by farmers to verify the individual forms before sending them. This way we implemented the privacy pattern of "Delayed Unrecoverable Action" [116, 146]. Since data can not be made private after publication, this is a crucial step to ensure that users get the chance to preview the released data before submitting it. To do this, the service requires the user to select the exported form file as well as the file with the disclosed values. The service then uses *form.io* to render a non-editable view of the individualized form.

8.6 Evaluation

To test how the participants from the agricultural domain would respond to a solution providing control over data visibility for business partners, we have evaluated the prototype we conceptualized after the prestudy in a think-aloud study, followed by semi-structured interviews and two questionnaires. The entire process comprising the creation of an evaluation guideline, recruitment, conduction of the evaluation, and data analysis and storage followed the guidelines of the ethics committee of

Technical University of Darmstadt. In the following, we will describe the evaluation steps in detail.

8.6.1 Method

Understanding the methodology of related works on the evaluation of individual visualizations is essential to identify a method that is best suited for the objectives of this paper. Authors from several research disciplines have already developed and evaluated visualizations with contributions ranging from purely descriptive comparisons of various approaches to complex usability studies of individual applications. After conducting a comprehensive comparison of visualizations in commercial community-centric visualization systems, Paci et al. [123] identified several requirements. Feth and Schmitt [137] also identify system requirements (for dashboards that are used internally in a company) but obtain these through qualitative workshop surveys. The testing of individual visualizing applications is dominated by analyses of usability, which is accomplished by assessing the effective execution of tasks [127, 129, 131, 136, 140]. In some cases, eye-tracking was applied simultaneously, for example by Bier et al. [131]. There, the System Usability Scale of the seminal work by Brooke [147] was used as an evaluation standard for usability. Overall, the usability analysis via task performance—possibly paired with think-aloud approaches or eye-tracking—seems to be the most suitable method to map important aspects of human-computer interaction.

8.6.1.1 Study Design

We conducted a think-aloud study to evaluate the prototype that was designed after the prestudy and to identify any obstacles the participants face when using a control matrix. This was followed by an interview and a questionnaire. The Electronic Supplementary Material provides an overview of the study flow for the evaluation (see A.1.2). The implemented prototype (see Sect. 8.5.2) itself was provided as a web app and accessible with a browser. The entire evaluation was conducted via video conference using *Zoom* with screen sharing enabled for the participants so that the investigator could see the actions of the participants and provide additional instructions when needed. The whole session was recorded. The questionnaire was filled out by the participants after the session to guarantee that the results are not biased by the presence of the interviewer.

Think-Aloud Testing. First, the participants received an introduction about the purpose and the procedure of the study and were asked for informed consent. While testing the prototype, participants were asked to comment on their actions and encouraged to express every thought. Then they received a description of a hypothetical scenario which can be found in the Electronic Supplementary Material (A.1.2). In the scenario, the user would be in the role of a farmer filling out a crop planning form providing information for a machinery ring, a service provider, and a seed supplier. The participant was then asked to load and fill out a form in the prototype and set the visibility functions of each data component. In the following step, the participant could download the data for each business partner and load the files to check how their business partner would see the form. Hence, the data with restricted access for the respective business partner would be left out in this view, and the participants could check whether the data in the form would be displayed correctly for each partner. In a third step, the hypothetical scenario was adjusted in that it would be one year later and the farmer would be planning the sowing for the same field. Now, the participants could load the previously filled out form, revise the data visibility, and then check the changes in the form again for each business partner.

Interviews. After the think-aloud test, interviews were conducted with the participants to inquire about a general impression of the prototype and to ask specifically about the perceived control the participants had. Here, we asked about the general perceived control in handling the prototype as well as about the control over who would receive which data. We asked the participants whether the forms for the business partners they checked were in line with their expectations. Also, we asked how the procedure would compare to the status quo within their own business practices. Lastly, we specifically asked for opinions on the control matrices.

Questionnaires. After the interviews, participants received a link to a questionnaire in the online survey tool *LimeSurvey*. While the small sample size was chosen for a qualitative approach and a quantitative analysis of the questionnaires would have limited meaningfulness statistically, we still wanted to inquire comparable data and get a more conclusive impression of the participants' attributes regarding technology affinity and their assessment of the prototype. To let them evaluate the usability of the prototype, we used the German version of the UEQ-S [148]. For their affinity for technologies, we used the German version of the ATI [149]. The questionnaires were presented in the mentioned order.

8.6.1.2 Participants

For recruitment, we reached out to our partners in the project HyServ to ask their organizations for participation. This way, we could interview members of two different machinery rings. Furthermore, we sent a request for participation to a federal counseling agency for agriculture and asked research assistants from [university] to participate. A compensation of 20€ was offered, but most participants took part in the evaluation without accepting money. The sample (N = 18) consisted of nine participants with an agricultural background (A1-A9) and nine additional participants with no experience in agriculture (R1-R9). We decided to gain general feedback on the usability of the system. The participants from agriculture mainly gave feedback on the possible integration into their daily work routine and use cases from this domain. While this is very useful feedback, sometimes the discussion circled around the question how to phrase some items or in which unit an item should be measured (e.g. the amount of seeds in tons, kilograms or bags). To gain insight from a more neutral perspective and to identify possible use cases others than agriculture, the second sub-sample was needed. Another reason was that all participants with agricultural background were male which could be a limitation, as the perspective of women could be underrepresented. To address this, we approached research assistants of our institution, to recruit a second sample with a more balanced gender-ratio. The research assistants are not directly subordinated to the leading authors. Their voluntary participation was credited as working hours, enabling compensation through the monthly remuneration. The agricultural sub-sample consisted of six farmers, of which four were familiar with the work of a machinery ring (A3, A7, A8, A9), one federal counselor for agriculture (A6), and two IT-experts with extensive experience in agricultural workflows due to their work (A4, A5). The age range in this group was 24 to 63, resulting in an average age of $M = 37.44$ years ($SD = 14.07$). The other sub-sample consisted of research assistants, of which four had a technical background (R1, R3, R5, R8), and five had a background in social sciences (R2, R4, R6, R7, R9). Five of the participants were female (R2, R5, R6, R7, R9) and four male (R1, R3, R4, R8). The mean age of this group was $M = 24.33$ ($SD = 1.41$), with the minimum being 22 and the maximum being 26. Detailed information on the participants can be found in table 8.3. This way, we could cover different backgrounds within our sample to assess how similar data sharing applications could be used even in contexts beyond digital agriculture. With regard to their affinity for technology, a Wilcoxon rank sum test showed no statistically significant difference between the two sub-samples regarding the ATI scores, which was contrary to our expectations ($W = 47$, $p = .593$). Our total sample scored an average of $M = 4.41$ ($SD = 0.87$) on the ATI scale, with only two participants scoring below the population mean of $M_{quota} = 3.58$ ($SD = 1.09$) as estimated by Franke, Attig, and Wessel [149]. A t-test showed the mean of our sample to be significantly larger than 3.58 ($T = 4.03$, $df = 17$, $p < .001$). This could indicate that participants in our sample have relatively high affinity for technology. However, it should be noted that a high score of affinity for technology does not necessarily imply a high IT proficiency. Nevertheless, a high affinity could positively bias the assessment of usability and be a limitation.

Table 8.3 Participants of the evaluation study

ID	Age	Gender	Background
A1	29	m	Farmer
A2	26	m	Farmer
A3	63	m	Farmer and machinery ring worker
A4	53	m	IT expert with experience in agriculture
A5	32	m	IT expert with experience in agriculture
A6	32	m	Federal counselor for agriculture
A7	50	m	Farmer and machinery ring worker
A8	28	m	Farmer and machinery ring worker
A9	24	m	Farmer and machinery ring worker
R1	23	m	Research assistant (technical background)
R2	24	f	Research assistant (social science background)
R3	22	m	Research assistant (technical background)
R4	24	m	Research assistant (social science background)
R5	26	f	Research assistant (technical background)
R6	24	f	Research assistant (social science background)
R7	24	f	Research assistant (social science background)
R8	26	m	Research assistant (technical background)
R9	26	f	Research assistant (social science background)

8.6.1.3 Data Analysis

The data analysis of the think-aloud data and the interviews was conducted in an analogous manner to the prestudy analysis described in subsection 8.4.1.3. The length of the interviews varied from 19 to 65 minutes, with an average of 40 minutes. The questionnaires were analyzed according to the instructions of their authors.

8.6.1.4 Limitations

The evaluation is subject to limitations regarding the sampling. Just as in the prestudy, gender- or age-based biases are possible. The regional distribution of the participants with an agricultural background within Germany is uneven. For a detailed discussion of these limitations, we refer the readers to Sect. 8.7.

8.6.2 Results

In this section, we present the results of the evaluations regarding transparent control over data. We refrain from stating feedback for single UI elements and wording, since this is not the focus of this paper. Therefore, we will leave out comments on the wording of agronomic terms in the labels of the form and similar feedback.

8.6.2.1 Usability

The control matrix was seen by most participants as an effective means of controlling their own data. All were able to understand how the views of the different parties were generated and that they could control this data flow. The results from the UEQ-S usability questionnaire confirm this assessment: The prototype as a whole scored high on the *Pragmatic Quality* scale with a mean of M = 1.48 (SD = 1.26). The authors of the test state that values above 0.8 represent a positive evaluation. On the *Hedonic Quality* scale, the mean score was M = .62 (SD = 0.62), indicating an average result. The mean scores and confidence intervals are shown in Table 8.4.

Table 8.4 Mean UEQ-S scores of the prototype

Scale	Mean	Std. Dev.	N	Confidence	Confidence interval*	
Pragmatic Quality	1.477	1.254	18	0.579	0.897	2.056
Hedonic Quality	0.616	1.094	18	0.505	0.110	1.121
Overall	1.042	1.009	18	0.466	0.575	1.508

* Confidence intervals for $\alpha = 0.05$ per scale

However, there was also criticism of the system: A1 criticized that farmers could prevent orders from being carried out by being too restrictive. Through unawareness, important information necessary for execution of tasks by third parties would be withheld which was also stated by A3 and A17. The term "may see" would need to be changed to "must see" in many cases. Other negative experiences were stated: R1, R2, and R8 found that a lot of scrolling was necessary and that the form was very long. A6 added, that the form should be reduced to the necessary minimum of items. R9 and A8 shared the impression, that the control matrix takes up a lot of space and impairs clarity. The overall experience of R8 was that it would be annoying to do this more than once and stated: *"You hope that you get through eventually"*. A similar experience was stated by A7, but with a more positive meaning:

(A7): *At first it seemed annoying, but the further you got in the form, the more sense it made since certain data really isn't needed by everyone.*

Some participants (R2, R4, R7) were surprised that every item could be managed separately, which was confusing at first. Furthermore, A8 brought up that farmers do not have the time it takes to manually configure the data access for everyone. In the case of multiple contractors, some would simply make separate disclosures for each contractor and set everything to visible. This view was also supported by the claim for pre-set values in the matrices by the participants A5, A6, A8, A9, and R5. In our prototype we gave a glance on these persistent settings by giving the opportunity to upload filled forms into the prototype. This way the participants would be able to reuse their settings from the first interaction with the tool. Furthermore, A8 and A9 stated that form-based workflows would only work if it would make further communication via phone obsolete. If additional phone calls could not be avoided with forms and the control matrix, using them would only be additional work for the farmers. As a solution, A9 proposed mobile support for these workflows so farmers could manage their data during meals or while waiting for the completion of other tasks, like the filling of a manure tank or supervising the satellite controlled machines in the field.

On the other hand, the control matrix was rated as easy to understand, even for first-time users (R3). A6, R3, and R7 found the two options easy to understand and approved the check, whether the rights are set correctly.

Another advantage of the control matrix was noted by A3. In agriculture, it is common that work orders are brokered by machinery rings. The control matrix would be an effective means to provide the machinery ring with the data needed for the mediation and, at the same time, to determine which requirements the performing parties would have to fulfill with regard to data processing and storage. The machinery ring could then select contractors who only need the data that a farmer wants to share. In this way, the control matrix supports standard farming practices and reduces the risk of non-completion of contracts due to unreleased information.

8.6.2.2 Default values for the control matrices

The default setting for usage rights was met with mixed feedback during the evaluation. A1 and R2 were initially confused and thought it was a demanded default that everything was set to "may not see." Further, R2 stated:

(R2): *Without preset, I would have been more concerned about each [right of use].*

However, other participants gave positive feedback. A6, R3, R4, R5, and R9 explicitly praised the default setting to "must not see." In addition, R5 stated that global settings for presetting would be useful.

> (R5): *A global "may not see" or "may see" for a given contractor would be useful to save work.*

Similar claims for "global" settings for each contractor were made by A8, R2, and R3. Such a procedure would lead to an inverted workflow: A contractor receives a permission for all data and, if necessary, individual rights are withdrawn. Such a procedure saves time, but could also lead to unwanted data disclosure.

Sharing data too sparingly can also lead to problems: A1 stated that jobs might fail if necessary data is not available. To avoid hindering job execution, R8 suggested that important data could be released by default. R9 has a similar view:

> (R9): *One could preset which data goes to the creator of the form, since he knows which data is absolutely required.*

This concept was also brought up by A4: Forms could be provided by service providers that contained pre-filled usage rights for essential parts of the order and would not be editable. A4 further commented that it would increase transparency if farmers were informed about why the use rights for some data are pre-filled and why that is necessary. This statement is particularly interesting because the word transparency was not mentioned previously during the interview and was introduced independently by A4 himself. In the study design, the term of transparency was not mentioned to the participants to avoid creating bias through nudging.

8.6.2.3 Sharing everything

Some of the participants without an agricultural background (R2, R4, R6) shared all data with everyone in the evaluation. This can be seen as an effect of the control paradox described in Sect. 8.2.2. Despite having the opportunity to control the data, the participants decided not to limit the access to their data. For example, R2 stated:

> (R2): *[It] is probably not relevant to everyone, but I am releasing it to everyone anyway.*

This does not necessarily suggest a lack of concern for data protection, but may also be due to the circumstances of the evaluation. R6 stated in the discussion:

(R6): *[I]f I had my own farm, I would have thought more about who should see what.*

These statements confirm what the agricultural test subjects also said: Accomplishing the task must be simple and fast. For farmers, digitization is extra work that they just want to get done. Form-based workflows are generally seen as a hassle compared to a quick phone call to the contractor. Also, passing on data to third parties is often tolerated in practice if there is a phone consultation beforehand. A9 reported based on his work experience for the machinery ring:

(A9): *Farmers do not care if everyone sees all the data, since most are not sensitive data. [...] The farmer is fine with the contractor calling in a third party to pass on the data directly [...] after consultation with the farmer.*

In the agricultural sample the rating of whether data is sensitive or not differs: While A9 wants additional consultations before sharing certain data, A8 does not consider most of the data in our scenario as sensitive, because most of the information could be obtained anyway by third parties. For example: The profit of farmers could be estimated by the size of their fields and the crop that is grown. Third parties could get both of these information by looking up the size of the cultivated land in official statistics and by observing the crop type on each field.

On the other hand, A9 confirmed that a specified and more restrictive disclosure is nevertheless useful:

(A9): *The other [parties] have the advantage of seeing only what they need to see.*

According to this, third parties even benefit from sparingly released data, as it facilitates the extraction of the relevant information from the order.

8.6.2.4 Advanced Rights of Use

In the interviews on the prototype evaluation, it was asked whether further rights of use for handling shared data should be included in the matrix. Examples of this would be the right to store the released data beyond the execution of the order or permission to share data with third parties.

From an agricultural perspective, A8 reported that the matrix should not be extended, as this would overwhelm users. This impression is shared by A8, A2, R2, and R3. Even if some participants consider the extension of the matrix by some specific columns useful (R1, R6, R7) or even necessary (R8), most of the participants agree that the system with two columns is particularly straightforward. Other solutions are suggested for the extended usage rights:

- Extended usage rights could only be granted globally for each contractor individually (A7).
- A third column could be added to the matrix, in which the use of the data can be specified via drop-down menu (R3).
- Additional columns could only appear when "may see data" is selected (R4).
- An individual definition of what exactly "may see data" should mean for the respective party could be added in a separate step (R5).

However, granting extended rights of use also has disadvantages, as R5 finds: Data can be disseminated unintentionally if the right to disseminate is granted by mistake. This shows, that the participants are aware of possible consequences of the control paradox, even if they are not familiar with the concept. According to R5, disclosure should always be prohibited. Sharing with subcontractors should only be allowed after explicit consultation with the data owner.

8.6.2.5 Differences between the sub-samples

In our evaluation we investigated the feedback of two sub-samples by including participants with agricultural background and research assistants without any agricultural background. The first sub-sample gave us detailed insights into possible integration of such data sharing applications. More feedback on the general usability of these systems without projecting them to the agricultural domain, was provided by the research assistants. We encountered the following differences between the sub-samples: The research assistants were more likely to be overwhelmed by the size of the form (R1, R2, R8, R9). This could be due the fact that they do not know the average amount of data that has to be shared in an agricultural scenario. The participants with an agricultural background seemed to be more familiar with the amount of data and were not surprised by the size of the form. On the other hand, the agricultural sub-sample was more likely to emphasize the need for time-efficient completion of the form (A5, A6, A8, A9). The familiarity with the frequency of data sharing situations in their working routine may be the reason for that. The third major difference exists regarding the pre-settings of the matrices. Many research assistants (R3, R4, R5, R9) praised the default setting of "not visible" to be privacy preserving. This was criticized by one farmer (A1) who sees the risk of a failing collaboration due to insufficient data disclosure. With regard of the usability questionnaire UEQ-S, we conducted t-Tests for the Pragmatic Quality ($T = 1.58$, $df = 15.21$, $p = .40$), Hedonic Quality ($T = -0.12$, $df = 15.73$, $p = .91$) and Overall Quality scales ($T = 0.84$, $df = 15.81$, $p = .83$), to test whether the groups differed significantly (p-values were adjusted using the Holm method [150]). None of the differences was statistically significant, even though we noticed a higher mean for the pragmatic quality in the sub-sample of research assistants (M = 1.93, SD = 1.33) compared to the agricultural experts (M = 1.03, SD = 1.06).

Although the prototype is still lacking some usability features and left out the transport implementation (see Fig. 8.3), the main goal of the evaluation could be fulfilled: We shed light on the willingness of users to control their data themselves, even if it comes at the cost of extra effort. Making controls transparent to users requires them to invest more time to configure the data disclosure, however, it also raises user awareness. By deciding who can access which data item, users reconsider whether anyone really needs to know the data. This way, data is withheld that otherwise might be disclosed thoughtlessly alongside other operational data to all participants. Even those who set everything to visible for all contractors had to reconsider whether they wanted to make their data accessible to all. This way, transparent controls contribute to more self-determined data disclosure and enable users to take responsibility for their data.

8.7 Discussion and Conclusion

In this paper, we investigated two research questions:

1. Which requirements exist for agricultural information systems to allow small enterprises to participate in cooperative work scenarios without becoming transparent for competitors?
2. (How) Can transparent controls contribute to better privacy behavior for farmers who are often time-constrained due to their daily work routine?

After conducting a prestudy with 52 participants from agriculture in Germany, we derived five challenges for data sharing applications in the domain of agriculture to address our first research question. The selection of challenges is based on our qualitative prestudy and does not claim to be exhaustive. However, these challenges point out important aspects of the technological development in the domain of agriculture. Unlike the IT expert users targeted for testing previous data sharing approaches [127, 129, 132], farmers have proven to have specific demands and expectations. We addressed these with requirements for IT solutions and presented considerations for the implementation of suitable tools. Based on these requirements, we implemented a prototype to investigate the second research question. By giving the participants direct control over data disclosure, we were able to verify findings from the prestudy. By means of our prototype, we evaluated that form-based access control with control matrices is a valid way to distribute data. This reinforces the finding of Tolone, Ahn,

Pai, and Hong [122] that, despite the old age of the approach, access control matrices remain effective and intuitive visualizations for data-sharing users. Addressing the shortcomings of access control matrices (see [122]) offers potential for future work. However, especially the agricultural participants pointed out that time constraints are a major factor in their office work. This often favors time-saving data disclosure in the trade-off between effort and control. In the following, we will discuss the results in more detail.

To address the *first challenge* of compatibility, we developed a form-based demonstrator utilizing JSON files. The distribution of these files can be done in a variety of ways, depending on implementation or background technologies. Sending them via email is possible, as well as uploading them to a cloud or attaching them to a smart contract on a blockchain. JSON files are machine-readable, which allows automated processing and thus is suitable for integration into existing management tools. Additionally, the use of JSON as a data exchange format fulfills the open source requirement stated in related work [69].

We addressed the *second challenge* of sector diversity by using a technology which allows the distribution of modified editors to generate JSON-based forms that could fit any purpose. By utilizing the control matrix, each user can decide which parts of the form should be available for whom. This way, the form can be used for different purposes and discloses as little information as required. In our evaluation, we presented a static form; however, via a drag and drop editor provided by *form.io* the customization of forms is easy to handle. In our scenario, the generation and customization would be done by expert users at the machinery rings or by contractors themselves.

The *third challenge* of media diversity was considered in the concept, but not fully implemented: In our evaluation, we did not support mobile versions as the interviews were conducted remotely during the COVID-19 pandemic. However, mobile support was not only a requirement from the prestudy, but was also mentioned during the evaluation: A9 suggested that farmers might fill out forms during meals or while in the field. Although our prototype had a responsive design to fit monitors of each size, full mobile support is still lacking and urgently needed. Using the *ionic* framework, we were able to build a responsive user interface. The extension to mobile support would only be a small step.

To investigate on the *fourth challenge* of transparency balance, we conducted a qualitative study with a prototype that gave users control over the data flow. We found that forms with control matrices are a proper measure to control the access to operational data within a job order. With the positive feedback on the prototype, being easy to understand for first-time users, we conclude, that the privacy pattern of "Direct Access to UI Components" [116, 145] was successfully implemented.

However, the additional control comes at the cost of time and effort spent configuring data access. We tried to minimize this effort by offering a pre-fill option to load previously completed forms to facilitate the use of follow-up contracts. With this feature, users just need to take the time to fill out the form once. Additionally, there are approaches that individualize the data sharing default, either through ex ante policy definitions or feedback-aware ex post mechanisms (e.g., machine learning). The work of Paci et al. offers a comprehensive overview of potential approaches to include an enhanced presetting for future work [123]. Overall, the feedback was mixed: The workflow was clear and well structured and the possibility to decide for each data object was appreciated. However, giving users control will take more time. Our study gave a peek on the willingness of users to make this trade-off. Furthermore, some participants stated the fear, that data could become disseminated unintentionally, if controls become too complex and users get confused and have no overview anymore. This could be seen as awareness for the effects of the control paradox. We tried to minimize this risk by setting all values to "must not see" by default according to the privacy pattern "Disable by Default" [116, 146]. This design decision was appreciated explicitly by many participants (R3, R4, R5,R9). Furthermore, we provided the user with an option to preview the data before sending in order to comply with the privacy pattern of "Delayed Unrecoverable Action" [116, 146]. This is a safeguard for the irreversible data disclosure. In the questionnaires, the pragmatic quality was rated good which leads to the conclusion that form-based access control is a promising approach. The hedonistic quality was rated neutral, which could be due to the fact that the prototype is still immature or the effort required in use. Another factor is that form-based tools are not yet widely used in agriculture. The amount of digital management is increasing, but many farmers still place orders by telephone. Compared with these calls, which can be made during work, form-based approaches are more complicated. On the other hand, we were able to prove that access control matrices with two usage rights ('is allowed to see data', 'must not see data') were easy to understand and did not lead to information overload. However, further investigation on the trade-off between effort and control is needed, for example with a qualitative follow-up study to validate our results. An important criterion for such a follow-up study is the investigation on the users' perception of their own possibilities to protect their data. In our study, we were able to gain insight into the fact that the sensitivity of business data was seen differently. While some farmers rated their data as very sensitive, others (e.g., A8) find the protection of this data tedious and unnecessarily time consuming, because third parties might find other ways to acquire this data anyway.

The *fifth requirement* is not yet met by our prototype. Due to the independence from a background technology, the tamper-proof store of the rights of use is lacking.

The users are able to store the JSON files themselves, but since the transport of data was out of scope, the ex-post traceability is not yet implemented.

The results are subject to *limitations*: (1) Our study has a focus on agricultural SMEs, because in this domain it is common that job orders include multiple actors and operational data needs to be distributed for highly digitized machines that require precise data. Nevertheless, these factors may also apply to other domains where SMEs are facing powerful up- and downstream players, which may gain advantages over them by access to operational data. (2) All interviewees from the agricultural sector (from both studies) run their farms in southwestern Germany, where the land structure differs fundamentally from eastern Germany due to the former division of Germany and former inheritance rules. Consequently, the study is not representative for agriculture in all of Germany, but readers should interpret it as a local picture of the western federal states in Germany. (3) Moreover, gender bias cannot be entirely ruled out, as the proportion of female participants for the agricultural samples in the study is well below 50 percent, potentially leading us to overlook challenges that affect women more than men. (4) Additionally, the sample included mostly young professionals, potentially reproducing negative stereotypes of the use of technology by older people [151]. (5) The qualitative method of the focus groups allows exploratory findings, but it cannot be used to test the hypotheses derived. (6) Also, in the prestudy, we could not always identify individual correlations between farm characteristics and challenges as it would have been possible in individual interviews. Despite these limitations, we have gained a detailed insight into farm processes and, in particular, data flows and corresponding challenges on which we can base the design of the data model.

For *future work*, we are planning to develop novel user-centered interventions that make the process of data disclosure even less time consuming. In the current prototype, we used control matrices since these are hardened constructs that were validated in previous research [122, 128]. We opted for this conventional control measure to prevent the bias that the control measure itself would not be usable and therefore the main focus of the evaluation on the effect of transparency on controls would not be accessible. Additionally, our findings could be used to construct a quantitative survey to validate our findings and get a more general impression of how farmers deal with the trade-off between effort and control.

To *conclude*, our contribution is threefold:

1. We conducted an empirical study with 52 German farmers to derive requirements for information systems in agriculture in order to minimize the threats of data misuse.

2. We then presented a data sharing model designed for decentralized data sharing systems such as blockchain ecosystems, which offers required features such as deletion periods and the definition of usage rights for data assets while preserving integrity and accountability to fight data misuse.
3. We developed a prototype with transparent controls to assist users in the data disclosure process for their business collaborations. This feature was then evaluated to estimate its impact on the working routine of non-expert users with time constraints such as farmers.

To summarize, farmers demand local processing of their data. Data ownership is a major issue in a domain where data sharing is a vital aspect of collaboration, while at the same time, however, this poses the risk of endangering the business continuity in case of data misuse. Form-based data disclosure with control matrices could assist users in minimizing the data they share with third parties while giving them transparency on who gets which data. However, users have to spend a certain amount of time and effort to adjust settings accordingly, which is not favored by users with tight time constraints.

Acknowledgments This work was supported by the German Federal Ministry for Education and Research (BMBF) in the project HyServ (01IS17030B), by funds of the German Government's Special Purpose Fund held at Landwirtschaftliche Rentenbank in the project Geobox-II and AgriRegio and by the Deutsche Forschungsgemeinschaft (DFG, German Research Foundation)—251805230/GRK 2050.

Paper D: Support Personas: A Concept for Tailored Support of Users of Privacy-Enhancing Technologies

9

Abstract

In many applications and websites people use in their everyday life, their privacy and data is threatened, e.g., by script tracking during browsing. Although researchers and companies have developed privacy-enhancing technologies (PETs), they are often difficult to use for lay users. In this paper, we conducted a literature review to classify users into different *support personas* based on their privacy competence and privacy concern. With developers of PETs in mind, support personas were envisioned to facilitate the customization of software according to the support needs of different users. In order to demonstrate the usefulness of support personas and based on workshop sessions with 15 participants, we designed a browser extension which supports users with the issue of script tracking by providing different user interfaces for different support personas. The following qualitative evaluation with 31 participants showed that the developed UI elements worked as intended for the different support personas. Therefore, we conclude the concept of support personas is useful in the development process of usable applications that enhance the privacy of the users while also educating them and thus potentially increasing their privacy literacy.

Original Publication K. Demuth, S. Linsner, T. Biselli, M.-A. Kaufhold, and C. Reuter, "Support Personas: A Concept for Tailored Support of Users of Privacy-Enhancing Technologies," Proceedings on Privacy Enhancing Technologies (PoPETs), *no. 4, pp. 797–817, 2024*, ISSN: *2299-0984*. DOI: https://doi.org/10.56553/popets-2024-0142. *[Online]. Available:* https://petsymposium.org/popets/2024/popets-2024-0142.pdf

Supplementary Information The online version contains supplementary material available at https://doi.org/doi.org/10.1007/978-3-658-47445-4_9.

S. Linsner, *Privacy Preserving Data Management*, Technology, Peace and Security I Technologie, Frieden und Sicherheit,
https://doi.org/10.1007/978-3-658-47445-4_9

9.1 Introduction

With the ongoing digitalization, many people around the world are using their computers and the internet for more and more purposes, ranging from private matters like web browsing to business applications like confidential meetings. Besides the threat of intentional cyberattacks [152], this extensive usage of the internet in everyday life results in people sharing a lot of personal information, intentionally and unintentionally with different companies. To mitigate this transfer of private information, security researchers have developed several privacy-enhancing technologies (PETs) [153, 154]. These technologies are intended to protect the user's data and prevent eavesdroppers or companies from acquiring information about the user. Often, the problem with these technologies is that their usage is too complicated for a majority of users [155]. This is, for example, one of the reasons why Pretty Good Privacy (PGP) is not widely adopted [156] and can only be used by people who have the required expertise and motivation to act accordingly [157]. Another problem is user reactance, i.e., users that mistrust a tool or security mechanism try not to use it, even though it would be more secure [23].

Research in the area of *usable privacy*—a research field at the intersection of human-computer interaction and privacy research [158]—tries to solve this problem. Existing work aims to identify user categories regarding their respective privacy attitude, e.g., by using users' knowledge and motivation [159], or distinct information cues users find important [160]. Further research has tried to find semantic relationships between users' traits, characteristics, and attitudes and their privacy profiles [161]. Another interesting research topic in the field of usable privacy is to investigate how users can be supported in their privacy decisions. Research has looked at the potential of nudging and soft paternalism [162], or how users can receive recommendations regarding their privacy settings [161, 163, 164]. In this paper, we combine the mentioned research topics and will classify users by their need for support while appropriating new PETs. In this case, support means the identification of the users' needs while using a tool and the suggestion of suitable settings based on these needs. For the classification of users, we use the concept of personas, which are archetypical users with different behavior, motivation, and goals [165]. Thus, it seems worthwhile to examine whether the identified user groups called "support personas" will differ from the "privacy personas" found by existing research [159].

Since most people use computers for browsing and thus are directly affected by the issue of script tracking, it was easily explainable in our workshop sessions and we used the topic to evaluate our concept of support personas in practice. For this purpose, we developed a browser extension prototype based on user workshops which makes script tracking more transparent to different user groups by implementing different user interfaces (UIs) tailored to the support personas. We chose this static instead of a dynamic approach to test the feasibility of our idea in a more controlled environment. Finally, we evaluated the extension to verify that it supports users with different backgrounds. In conclusion, this paper answers the following research question in the context of script tracking: **How can developers of privacy-enhancing technologies provide users with tailored support?**

Based on our reviewed literature (Sect. 9.2), the main contribution that is provided is the methodological contribution of the novel concept of support personas, which can be used to tailor the support for users trying to appropriate new technology (Sect. 9.3). As an empirical contribution, our workshops shed light on the UI requirements of user groups with different privacy concerns and competencies (Sect. 9.4). Finally, we provide an artifact contribution by the participatory design of our demonstrator PRIVACYASSIST, which is a browser extension making script tracking more visible to the user (Sect. 9.5). Based on the evaluation of its user experience and user support capabilities (Sect. 9.6), we discuss the core contributions of our study (Sect. 9.7) and conclude the paper (Sect. 9.8).

9.2 Related Work

For the design of support personas, we reviewed literature on the appropriate categorization of users, different modes of user support, and the foundations of our intended application domain (i.e., privacy and script tracking in browsers), which is summarized in the following subsections.

9.2.1 Categorization of Users

Personas are an important concept to identify users' needs and design applications for distinct user groups specifically [40, 166]. Previous research described different approaches how personas can be developed. One possibility is to use quantitative data and statistical methods in order to identify relevant personas [167, 168]. Another possibility is to use only qualitative data [169].

For privacy applications, previous research developed privacy personas to categorize users regarding their privacy behaviour and attitudes. One of the first categorizations of users is the model of Westin [170], which differentiates users regarding their privacy concern. He identified three groups of users, i.e., the Unconcerned, the Fundamentalists, and the Pragmatic Majority. The Unconcerned are users who trust organizations to collect their personal data. The Fundamentalists are the opposite. They do not trust organizations to handle their personal data. Regarding the privacy paradox, they often prioritize their privacy over benefits they would receive for sharing data. According to Westin, the majority of users are Pragmatists. They weigh their privacy with the benefits they would receive for sharing personal data and are willing to share them if they receive an adequate benefit. To classify users according to those categories, Biselli, Steinbrink, Herbert, Schmidbauer-Wolf, and Reuter [171] developed a questionnaire. The questionnaire contains questions regarding the privacy knowledge and the privacy behavior of users and is used to shed light on the general methodological difficulties of accurately classifying users.

Another approach that obtained approval in the research community was developed by Morton and Sasse [160]. They categorized users into five groups with regard to both their privacy knowledge and behavior: Information Controllers, Security Concerned, Benefit Seekers, Crowd Followers, and Organizational Assurance Seekers. This categorization was also used by Dupree, Devries, Berry, and Lank [159]. They investigated the privacy motivation and knowledge of users and then examined what clusters developed. Thereby, they confirmed the five clusters found by Morton and Sasse [160]. In conclusion, research has focused on classifying users into different so-called privacy personas. Primarily, this categorization reveals users' attitudes towards privacy and the protection of their data. To the best of our knowledge, there are no works that have classified users regarding the support they need and want when working with PETs.

9.2.2 Support of Users

In order to elaborate what kind of support users want, we reviewed related work in this research area. For secure systems, Holmström [172] identified a growing need for usable and comprehensible security solutions. Therefore, she developed a security concept to support users with their security decisions and issues. As essential steps when designing a UI, she identified determining the user's security awareness, knowledge, and needs. As Ray, FlynnWolf, and Aviv [173] found out, people in different age groups have different privacy concerns, highlighting the necessity of addressing individual users differently. This is also important since users with different backgrounds have differing privacy knowledge and behaviors [174, 175]. Especially lay users are often unaware of existing privacy risks [176].

With a focus on the UI, MacDonald and Smith [177] developed design guidelines for interfaces that raise users' awareness of application security. According to them, the design of the interface is crucial for the security of a system, since the computer depends on the information conveyed to the user. Without correct or adequate information, many problems can arise, e.g., confusion or an increased frequency of errors. A good interface design can reduce the risk for these problems and thus increase the system security too. The developed guidelines include the usability, the path of least resistance, the recoverability and revocability of actions, and the explicit user authorization. Other principles [178] imply that the UI should provide elements to inform the user about the current security state of the system in a comprehensible way. Since the user's attention is limited and security is usually not their priority, the UI design should be frugal and simple, while providing the information required to make the right decisions.

However, challenges exist even in software designed to support users. Acquisti, Adjerid, Balebako, Brandimarte, Cranor, Komanduri, Leon, Sadeh, Schaub, Sleeper, *et al.* [162] studied problems in privacy and security decision making. The identified problems are incomplete and asymmetric information presented to users, heuristics and bounded rationality, as well as different cognitive and behavioral biases. Another problem found by Wu, Zhang, Ma,Wen, Xia, Paris, Nepal, and Xiang [45] is that texts about security in applications often contain various technical terms that are not understood by many users. In an experiment they verified the success of using explanations in pop-ups to help users understand such texts. It was also found that there is a big difference between the developers' assumption on user privacy expectations and the actual privacy expectations of users [179].

A concrete application to support users was developed by Barth, Ionita, De Jong, Hartel, and Junger [46]. They built a privacy rating aimed at helping users to understand how an online service handles their data. The privacy rating was realized by calculating a score which was well accepted by users, including lay users. However, the authors identified problems regarding incomprehensible terms, as users required background knowledge to understand them. Namara, Sloan, and Knijnenburg [180] tried different concepts to engage users to interact with privacy features on social networking sites. They found that an automated system comes with the most secure settings for users but a suggestion system best facilitates interaction between users and the system. In summary, previous research examined which support users need when using PETs and what problems can arise in the context of user's data privacy and security. Nonetheless, there is still a demand for applications that adjust their support to the user's needs in order to support the individual user and not just the main target group of a software, e.g., by helping them to understand security terms.

9.2.3 Privacy in the Browser

For different browsers, various browser extensions exist which implement different solutions on how users can prevent online tracking. Mathur, Vitak, Narayanan, and Chetty [181] examined the use of such extensions. They present three main findings. First, they found that most internet users only have basic knowledge about online tracking. Second, they observed that every extension has one specific primary use case. Third, they revealed that extensions do not break websites very often, i.e., block too many website elements so that the website does not work anymore. Besides browser extensions, browsers also integrate functions to inform users about the security of a website. One of those functions is the padlock icon, which is displayed in the address bar to signalize an encrypted connection. In a study by Zezschwitz, Chen, and Stark [182], it was discovered that most users did not know the meaning of this icon. The authors propose that a secondary UI can communicate the meaning of icons or the security of a website or connection more clearly. A similar problem was identified by Mehrnezhad, Coopamootoo, and Toreini [183], who found that users learned the most about privacy and security by conducting their own research or through friends and family and not by using PETs. Additionally, users often do not know how to effectively control the privacy practices of websites when browsing [184].

Farke, Balash, Golla, Dürmuth, and Aviv [185] examined which impact privacy dashboards, i.e., windows showing the collected private information of the user, have on the end user. They tested this with the Google My Activity dashboard. After using the dashboard, most participants were less concerned about data collection and some stated that they would change a few settings in the dashboard. On the one hand, the researchers highlighted that privacy transparency tools would increase users trust and improve their privacy perception. On the other hand, they doubted whether these tools encouraged users to view the collected data and change settings. Weinshel, Wei, Mondal, Choi, Shan, Dolin, Mazurek, and Ur [186] developed a browser extension to raise awareness for web tracking. They implemented different interfaces showing different information to the user and evaluated those.

To conclude this section, recent literature shows that users of privacy extensions for browsers often have little knowledge about privacy issues and that a secondary UI providing more details about the security of a website or collected data and thus educates the users would be beneficial. This secondary UI can be provided in form of an extension.

9.2.4 Research Gap

The literature shows that a categorization of users regarding the required support when working with PETs currently does not exist. This paper develops such a categorization by introducing support personas. The concept is then tested in a development process for an application with UI components tailored to the individual user and the assigned support persona. The demonstrator is a browser extension that presents users with more details regarding script tracking and combines approaches from different existing applications. This is relevant since literature shows that existing applications only serve one primary purpose, e.g., a privacy score [181]. During the design of the application, we adhered to the design guidelines established by Spero and Biddle [178]. The use case of script tracking allows us to implement different visualization techniques which can also be used to visualize other tracking techniques. This is shown by existing applications which use some of the techniques to visualize more tracking techniques like trackers or encryption status. For our use case as a demonstrator, the limitation on scripts is a limitation but still demonstrates the use case since the UI elements and their meaning is the same except for the calculation in the background which uses only scripts in our case. Nevertheless, due to the complexity of tracking, a browser extension for more tracking techniques like cookies or fingerprinting would result in a more complex user interface.

The concept of support personas is related to the concept of privacy personas developed by different researchers, e.g., [159, 160]. These personas were largely based on the users' privacy behavior. It was shown [187] that users' reported privacy behavior and real privacy behavior differ. Therefore, we needed to create the support personas to provide users not only with the information they think they need but with the information and support they really need based on their objective competencies.

A research work closely related to our demonstrator is the one by Weinshel, Wei, Mondal, Choi, Shan, Dolin, Mazurek, and Ur [186] who also developed a browser extension for web tracking. Although they tested various interfaces showing different information in a field study, they did not develop different interfaces for specific user groups, in contrast to our work. Another closely related work is the research by Mathur, Vitak, Narayanan, and Chetty [181] who examined the relationship between the privacy knowledge of users and the usage of browser-based blocking extensions. Our work builds on research by Mathur, Vitak, Narayanan, and Chetty [181], as they found that users lack knowledge about online tracking and that every extension only serves one purpose, such as a privacy score or detailed information. In addition, we aim to improve user privacy competence and combine different existing extensions in our extension.

9.3 Support Persona Items

A persona is an archetypical user consisting of attitudes, motivation, behavior, goals, and more [40]. Schneidewind, Hörold, Mayas, Krömker, Falke, and Pucklitsch [40] showed that personas can be used to solve different challenges that arise for requirements engineering. They propose that the personas in a given scenario should be developed before determining the requirements for a project. This way, the personas increase the user focus and awareness and can contribute to a better understanding of future users' behavior and needs [166]. In this work, we will adhere to this and develop the support personas before the implementation. The support personas classify users regarding the support they need while appropriating new PETs. Since this paper focuses on privacy-enhancing technologies, we used the two dimensions of *privacy competence* and *privacy concern* to create support personas. Similarly to previous works clustering users [159, 170], we chose to divide each item into three groups (*low, medium, high*). The dimension *privacy behavior* was not chosen, because it was shown that the stated privacy behavior often differs from the actual privacy behavior [188]. The *used privacy practices* are another potential dimension. That was not chosen because, while users may already use PETs, it is possible that they do not know how to use them correctly. This is due to the unavoidable different user groups which use the same application.

9.3.1 Privacy Competence

Holmström [172] found, it is essential to recognize the user's knowledge of security issues for the development of a user interface for PETs. This way, it can be ensured that only user-understandable information is displayed [45]. Furthermore, more experienced users may not require some explanations, as this could disrupt their work. Additionally, a customized user interface prevents confusion by only displaying information relevant to a specific user. This can enhance the entire system's security [177]. For UIs in general, items like tech literacy could be used to determine the knowledge of users. But since support personas should be used for PETs, we decided to choose a more privacy focused item. Also, Crossler and Bélanger [189] observed that the users' privacy competence determines their use of privacy-protective settings. Thus, we used the user's privacy competence as the first factor to create the support personas.

In order to classify users regarding their privacy competence, we need to establish a measure. For this purpose, we chose the questionnaire 'Online Privacy Literacy Scale' (*OPLIS*) by Masur, Teutsch, and Trepte [190]. This questionnaire consists of

20 questions regarding the user's online privacy competence[1] and is shown in the Electronic Supplementary Material (A.2.3.1). Although the *OPLIS* scale does not exclusively measure knowledge of tracking technologies, it includes items related to cookies and other similar technologies. Moreover, having a comprehensive understanding of online privacy is crucial for making informed decisions. Therefore, it is suitable for our purpose of identifying the user's privacy competence. Using this questionnaire, we built three user groups according to their privacy competence. A user can achieve a maximum of 20 points by answering all 20 questions correctly. Since the authors [190] attached norm tables to their work, we will use the norm table for the whole population determining how many people score how many points on average as an indicator for our groups. The first group we built is the one with the least privacy competence. Users in this group achieve ten or less points in the *OPLIS* questionnaire and could be users that want to protect their privacy but lack the competence to do so. Therefore, they may need more explanations and easier language in the application. Users who achieve between eleven and 14 points are in the second group, which comprises users with average privacy competence. This user group will probably be the largest one comprising both people that want to protect their privacy actively by learning more about the topic and people that do not want to learn new things and want the application to just work. The last group with the most privacy competence consists of users achieving 15 points or more. This group includes the expert users that may want the most configurable options and information. We established the boundaries between the groups using the mentioned norm table for the whole population. According to the norm table, each group should represent one third of the whole population regarding privacy competence.

9.3.2 Privacy Concern

Furthermore, Holmström [172] identified the user's awareness and security needs as important factors when creating a UI. A definition of information privacy concern was developed by Malhotra, Kim, and Agarwal [191], who defined it as an individual's subjective views of fairness within the context of information privacy. This aspect is of particular interest for our support personas, as users with differing attitudes on privacy need to be addressed differently. For example, a user worried about privacy does not need to be motivated to use PETs but only needs to know what and how to do it [192]. In contrast, users without worries about privacy do not necessarily want privacy warnings to disrupt their work flow and probably have

---

[1] https://www.oplis.de/index_eng.html

increased risk taking behaviour [193]. Brough and Martin [194] underline that both privacy concern and competence are important for the privacy behavior of the user. Thus, privacy concern is the second factor we looked at when creating the support personas.

As a scale for privacy concern, we chose the "Internet Users' Information Privacy Concerns" (*IUIPC-8*) by Groß [195], which is based on *IUIPC-10* by Malhotra, Kim, and Agarwal [191]. We chose this scale since it is a well established instrument to quantify privacy concern [196]. The questionnaire consists of eight items (cf. Electronic Supplementary Material A.2.3.3). These belong to the dimensions control, awareness, and collection, which were shown to be relevant dimensions for privacy concerns. The questions are answered by using a Likert 7-point scale ranging from 1="strongly disagree" to 7="strongly agree". A person answering all questions with "7" is considered to have very high privacy concerns. A person answering all questions with "1" is considered to have nearly no privacy concerns. Building on this, we develop three user groups: One consisting of users with low privacy concerns (total score of 19 or less), one with medium privacy concerns (total score between 20 and 38), and one with high privacy concerns (total score of over 39). In the first group, there might be users who are not concerned about their privacy and therefore not willing to accept any additional overhead on their main task to protect their privacy. The second group might comprise of people who never thought about taking care of their privacy but are willing to learn how to protect it. The last group of users care about their privacy and are also willing to accept additional overhead in order to protect it.

9.4 Participatory Design Study

Our implementation aims to provide users with information about script tracking on websites. Although some browser extensions providing information about script tracking already exist, none of these provide different interfaces for different user groups. Thus, we want to build a browser extension that adapts to its user. In contrast to existing "simple" and "advanced" views (i.e., providing only a general overview versus providing more information and configuration options), our different views will be adjusted to the needs of a user presenting necessary explanations to lay users and more details to sophisticated users. To develop such an extension and the support personas resulting from the developed items (cf. Sect. 9.3), we used a participatory design approach. Participatory design has proven to be important in software and product design in order to consider special characteristics of the future user, either to meet the requirements of a special user group like the elderly [32], or to generally involve different stakeholders in the design process [197]. Specifically, we organized a series of workshops to receive input from different stakeholders, i.e., potential users of the extension. The demographic data of the workshops, surveyed via a form that was filled out by the participants at the beginning of every workshop, can be seen in the Electronic Supplementary Material (2.1.1). Although most of our participants were young and well educated, the participants were mixed in terms of knowledge of tracking technologies in browsers. Especially in workshop 2, most of the participants were unfamiliar with the technology behind the examined concept of script tracking, have not used browser extensions to visualize tracking, and needed an introduction. The methodology of the workshops is described in the following section.

9.4.1 Methodology

We conducted three workshops to develop the concrete support personas and the concept of our implementation. Before conducting the workshops, we obtained an IRB approval (EK 16/2022) by the ethics committee at our university. Each workshop was held with five German students from different fields of study, who signed an informed consent form informing them about the workshop and the collected and processed data. The students were recruited through a convenience sampling method from personal and professional networks of the researchers without special selection criteria and were not financially compensated. Therefore, they were from the same university except for two students who study at other universities. The workshops were held in person, had an average duration of 79 minutes, were neither video- nor audio-recorded, and comprised a (1) motivation of the topic, (2) presentation of known applications, (3) work phase, and (4) conclusion phase.

9.4.1.1 Motivation of the Topic

At the beginning of each workshop, the topic and our motivation was presented. This was inspired by the introductory section of this paper (cf. Sect. 9.1). Additionally, a short introduction to script tracking was presented to explain the concept to the participants. This way, we ensured that every participant knew the basics of the topic and could participate during the workshop. To ensure comparability between the workshops, we followed a script (cf. Electronic Supplementary Material A.2.1.2). We focused only on script tracking instead of other tracking technologies, because otherwise the cognitive load of the participants would possibly be too high to focus on the support personas and the design of the application. Additionally, this limitation does not influence our implementation since it is only a demonstrator which shows how support personas can be implemented in a design and implementation process.

9.4.1.2 Presentation of Known Applications

After the introduction, a few applications offering different UI solutions on how to visualize script tracking and other tracking elements to the user were presented. These applications were Ghostery, Disconnect, DuckDuckGo Privacy Essentials, NoScript, and UMatrix.

Ghostery Ghostery (see Electronic Supplementary Material A.2.2.1) shows the number of tracking-related requests as a number in the icon on the extension bar. After clicking on the icon, the user can see the different categories of the identified trackers. Additionally, statistics about the number of blocked trackers, the number of changed requests, and the loading time of the site are presented. In the extended view, the user can see which requests were classified in which category and the status of every request (i.e., blocked or unblocked). Additionally, the user can block or unblock single requests in this view.

Disconnect Disconnect (see Electronic Supplementary Material A.2.2.2) also categorizes the found tracking-related requests but in addition, shows how many trackers were found from Facebook, Google, and Twitter. Additionally, in the normal view the user is shown how much faster the website loaded and how much bandwidth was saved by using Disconnect. After clicking on the different categories, the user sees the trackers that were blocked in this category. Here, the user can block or unblock individual or all trackers in a category. Finally, Disconnect offers a view where the user can see the trackers visualized as a graph showing the websites the trackers were loaded from. Websites the user has already visited are highlighted.

DuckDuckGo Privacy Essentials DuckDuckGo Privacy Essentials (see Electronic Supplementary Material A.2.2.3)) assigns a privacy score to every website. This score rates a website's trustworthiness. The score improves when the connection to the website is encrypted, fewer trackers are used by the website, fewer known tracking networks are found on the website, or the privacy policies are known. It is also shown how much the score is enhanced by using the extension to block some trackers or enforce the encryption of the connection.

NoScript NoScript (see Electronic Supplementary Material A.2.2.4) is an extension that blocks all scripts by default. This can impair the proper functioning of websites the user visits for the first time. The user then has to explore manually which scripts are required for the website to function and unblock scripts accordingly.

uMatrix uMatrix (see Electronic Supplementary Material A.2.2.5) analyzes all building blocks of a website and presents them in a matrix. The rows show the different domains the building blocks are loaded from. The columns tell the user which kind of content is loaded (e.g., a picture, a cookie, or a script). The user is also shown which sources are blocked and which websites as script sources are considered not to be relevant for the website to work.

9.4.1.3 Work Phase

To allow the participants to think freely about concepts of implementation, they were not informed about the results of Sect. 9.3, i.e., that we plan to divide the users into three groups per dimension. They were only told that the different user groups will be based on the privacy competence and privacy concern of the users. Thus, all workshop groups considered the first step in the work phase to be to think about possible user groups arising from these two characteristics. After identifying those groups, the participants thought about UI elements suitable for each user group. The results of this phase were noted on a white board and visible for every participant during the entire workshop. This way, the results could be refined and used as a basis for new ideas.

9.4.1.4 Conclusion Phase

At the end of every workshop, we summarized the results. This gave the participants the chance to clarify misunderstood ideas or add important details.

9.4.2 Results

9.4.2.1 First Workshop

The first workshop (90 minutes) was held with five male participants studying computer science or electrical engineering on 07/15/22. After the introduction and motivation of the topic, the participants began to develop user groups based on the users' privacy competence and privacy concern. The first idea was to distinguish between low and high privacy competence and low and high privacy concern, leading to four user groups. After some discussion, two participants mentioned that they do not entirely identify themselves with any of the groups. Thus, a new idea to implement a medium category for both items was developed. Now all participants found themselves in one category.

After establishing this classification of users, the participants discussed whether some of the categories could be merged in the implementation since users of different categories share the same requirements for the UI. The discussion ended with the result that the participants would implement three different views in the UI depending primarily on the person's privacy competence and less on the privacy concern. This was concluded because the privacy competence would be much more important for determining the level of support given by an application, while the privacy concern would only determine how the user uses the given setting options.

In the last step, the UI elements for the three views were developed. For the first group, i.e., users with low privacy competence, the view should be designed as simple as possible. The participants found a score, similar to that from DuckDuckGo Privacy Essentials, to be helpful. The extension should also tell the user how many scripts were blocked. Additionally, the pop-up should contain some tips for education and knowledge expansion. In the pop-up, a button to turn off the script blocking on the current website should be implemented. This button should be large enough to be easily noticed by users with less experience. The UI view for users with a low privacy competence should serve as base for the other views. The view for users with medium privacy competence extends this view by introducing different categories of scripts. In the workshop, this classification was not specified but was later implemented by using the categorization of third-party scripts and others developed in the third workshop. Users in this category can decide to block or unblock all scripts of a category. The third and most advanced view for users with high privacy competence lists the individual scripts in all categories and allows the user to block or unblock single scripts. Further, it is shown when scripts from one origin are loaded on many sites visited by the user.

The participants also pointed out the importance of the user's ability to customize the interface. For example, a user with low or medium privacy competence should be able to change the view to the most advanced view if wanted. This setting option must be implemented for all three views.

9.4.2.2 Second Workshop

The second workshop (79 minutes) was held with five participants, three of them female and two male on 07/28/22. They were students from different universities studying psychology, computer science, finance, or medicine. After the introduction and motivation, they started to identify four relevant user groups based on users' privacy competence and privacy concern. The first group consists of users with low privacy competence and low privacy concern, the second group of users with low privacy competence but high privacy concern, the third group of users with medium competence and medium concern and the last one of users with high privacy competence and high privacy concern.

Thereafter, the participants were told to find suitable UI elements for different user groups. During this task, they decided that the requirements for a UI should be linked to the privacy competence of the users. Therefore, users of the first two groups were assigned to the same UI. Thus, three UIs were developed.

The first UI for users with low privacy competence contains a score similar to DuckDuckGo Privacy Essentials. Additionally, the users should be presented with the categories of found scripts similar to the regular view of Ghostery. In this view, it should be possible to turn off the blocking of all scripts in case a website does not work. Additionally, the user should get more information and explanations on scripts when clicking on little information buttons ('i'). The second view for users with medium privacy competence should contain a more elaborated score, i.e., with information about its calculation. In addition to the simple view, this extended view should show which script belongs to which category. The most detailed view for users with high privacy competence should contain a matrix similar to the one of uMatrix. The participants found this to be an efficient way to provide the most detail. For users who want even more information, the participants found a graph view like the one in Disconnect helpful. This graph view should be accessible over a button in the detailed view.

The participants pointed out that a user should be able to switch between different views. Further, when the user changes settings in the extended or detailed view, the on/off button in the simple view should adjust and show a custom setting. This way, the user sees that a property was changed when returning to the simple view.

9.4.2.3 Third Workshop

The third workshop (69 minutes) was held with five participants studying computer science, electrical engineering and information technology, business informatics, and computational engineering on 08/10/22. After the obligatory introduction and motivation of the topic, the participants of the workshop were asked to identify user groups based on users' privacy competence and privacy awareness. Their first thought was to draw a cartesian coordinate system with privacy competence on the x-axis and privacy concern on the y-axis. With this coordinate system on the white board, they started to think about possible user groups inside this system. The first approach mentioned was to divide every axis into two groups, thus having four groups. After a few discussions, the participants concluded that four groups are not enough since 'people in the middle', i.e., with medium privacy competence and concern, would easily fall into one of the extreme groups, e.g., with high privacy

Table 9.1 Differences and similarities (highlighted in bold) between the workshops

First Workshop	Second Workshop	Third Workshop
• **Three UIs based on privacy competence**	• **Three UIs based on privacy competence**	• **Three UIs based on privacy competence**
• **Possibility to change views**	• **Possibility to change views**	• **Possibility to change views**
• **Three categories per item**	• Four user groups: low competence and concern, low competence and high concern, medium competence and concern, high competence and concern	• **Three categories per item**
• Privacy concern determines how user uses settings		• Privacy concern as a measure to set default settings
• First UI: **Turn-off button, score**, Number of blocked scripts, tips for education and knowledge expansion	• First UI: **Turn-off button, information and explanations, score**, script categories	• First UI: **Turn-off button, information and explanations**, score information to compare security of websites
• Second UI: **Show scripts in categories, (un-)block by category**, Extend first UI	• Second UI: **Show scripts in categories**, score with calculation information	• Second UI: Score, **(un-)block by category**, number of scripts
• Third UI: List all scripts sorted by category, (un-)block single scripts, show often appearing scripts	• Third UI: **Matrix view showing building blocks**, graph view of tracking elements	• Third UI: **Matrix view showing building blocks**, as much information as possible

competence and concern. Thus, they decided to divide every axis into three groups (low, medium, high) resulting in nine different more fine-grained user groups.

After identifying the user groups, the participants developed the UI of the browser extension. Their approach was to use the privacy competence as a measure to decide which UI is shown to the user. This strategy results in three different UIs, called simple, standard, and expert. The participants mentioned that the user should be able to switch between the interfaces. The privacy concern, on the other hand, should be used to set the default values of the extension, i.e., how many scripts get blocked. For users with a low concern, the extension should block less scripts so that every website works. With increasing concern, more scripts should be blocked. The appearance of the three UIs is described in the following.

The first 'simple' UI should mainly consist of a privacy score like the one by DuckDuckGo. The participants mentioned the idea to extend the privacy score with information on how secure the site is compared to other sites the user visited. This should make the very abstract score more concrete and tangible to the user. Additionally, the simple view should feature a button to disable the extension or block more scripts. The default setting of this slider should be based on the user's privacy concern. Additional information and explanations could be given to the user by using little question marks. The second user interface called 'standard' should contain the privacy score from the 'simple' interface. In addition, the number of scripts should be shown. The slider from the 'simple' interface should also be extended by providing more information about the blocked scripts, e.g., that all external, i.e., third-party, scripts are blocked or that external and internal, i.e., not third-party, scripts are blocked. The third 'expert' user interface should provide the user with as much detailed information as possible. For this, the participants of the workshop found the matrix representation of the website from uMatrix suitable. An interesting idea mentioned was giving the user the possibility to view the script to be loaded so that an expert could look at it to find out what it does and decide whether it is important. The participants also suggested to use buttons included in the 'expert' view to show the user as much information and statistics as possible upon request.

9.4.3 Summary

In this section, we will summarize the workshops. The differences and similarities between the workshops can be found in Table 9.1.

9.4.3.1 Support Personas

A common result of all workshops was to use three gradations on the items privacy competence and privacy concern. This result hardens the findings of our literature work (Sect. 9.3) by confirming the identified support personas in a qualitative setting. The first and third workshop worked with all nine resulting user groups, while the second one only identified four user groups. Still, all workshops proposed three different views based on the privacy competence of the user. The privacy concern was only mentioned by two workshops as a mean to set the default settings or describe how the user will use the settings.

For the support personas, this means that the privacy competences of future users are of special importance during the development process. The workshops showed that the interface of PETs should be adjusted to the privacy competence of the user, so that every user gets understandable and relevant information. The privacy concern gets important later in the development process of PETs where the default settings for users are determined. At this stage, users with a higher privacy concern might accept more drawbacks for an improved privacy than users with a lower privacy concern.

9.4.3.2 Browser Extension

According to the workshops, the simplest view should contain a privacy score. Additionally, little question marks should provide explanations when the user clicks on them. This is important for users who want to acquire more knowledge. One workshop concluded that the simplest view should also present the number of blocked scripts and the script categories. Since the other workshops determined that the categories should be mentioned in the second view, we will only show the number of found scripts in the first view. Finally, a button should be implemented to deactivate the extension for a website in case it does not work. According to two workshops, the second view should, besides the privacy score, also show the categories of scripts and how many scripts were found in each category. The most extensive view should present the number of found scripts in every category together with the individual scripts and as much information as possible. In the workshops, the matrix representation was found to be most useful for the primary window. After clicking on different buttons, the user should be able to access even more information, e.g., by using a graph view like the one from Disconnect.

9.5 Design of PrivacyAssist

We built the PrivacyAssist extension for the Firefox web browser[2]. For development, we used the integrated development environment (IDE) WebStorm[3] and tested it by using the Firefox development tools. In the following subsections, we discuss and present the concept, first prototype and final version of the browser extension.

[2] https://www.mozilla.org/firefox/

[3] https://www.jetbrains.com/webstorm/

Table 9.2 Overview of the identified and addressed challenges

Challenge	Literature / Workshop Findings	Solution
Give the user confidence in own capabilities	[159]	Design of different UIs based on the user's privacy competence
Prevention of user frustration	[177], W3	By default, block only external scripts as this should not break websites
Easy to use and understandable UI	[162, 177, 178]	Design different UIs for users with different privacy competence and provide help and additional explanations
Provide more detailed information to experts	W1–W3	Design a view with very detailed information for users with high privacy competence
Freedom of choice	W1–W3	Provide the user the possibility to change views

9.5.1 Concept of the Browser Extension

The concept is based on the studied literature and the workshops (cf. Table 9.1). It addresses the identified challenges, which are illustrated in Table 9.2. In Sect. 9.3, we highlighted that the user must have the feeling that their taken actions will be sufficient to improve their own privacy. In our implementation, we take this into account by offering different UIs based on the respective privacy competence which provide users with comprehensible information and explanations. This solution was also suggested in our workshops. Since we discovered that user frustration can arise from poor interface design, we try to offer users with the most suitable interface by default after an evaluation of the respective privacy competence. Moreover, we ensure that a website can function properly and is not impaired by only blocking third-party scripts by default. The user can change this setting for every website. An easy to use and comprehensible interface also contributes to more secure decisions by users. We achieve this by providing a very simple interface especially for users with less privacy competence. The simple interface only shows the privacy score, which is easily recognizable in the menu bar. When the user clicks on the icon, the privacy score and some easily comprehensible information about it is shown, e.g., how it is calculated. Additionally, small buttons with question marks are used to provide additional help and explanations.

With increasing privacy competence, the user can handle more detailed information. In the second view, the user is shown the different categories of scripts and has the option to block an entire category. The privacy score is also shown in this view. The most extensive view shows all found scripts, sorted by category and has the option to block individual scripts. In the future, this view can be extended so that the user is provided with more statistics and information, e.g., a graph view like the one from Disconnect.

In the workshops, it was concluded that the user should be able to change views upon request. Hence, we implement this functionality by using different tabs in the browser extension. Additionally, it was mentioned that the title of the different views should not give the user the feeling of having insufficient knowledge to use other views. Thus, the views are labeled "Standard", "Extended", and "Detailed".

9.5.2 First Prototype

The first step of the process was to build a prototype to extract information about the scripts from a loaded website and display it in the pop-up window of the extension. Since the source of a script can be used to assess its function, we decided to extract the sources and display them in a list. A browser extension consists of different parts. The most prominent part is the pop-up. It expands and is only active when the icon of the extension in the top right corner of the browser is clicked and is built like a website, consisting of an HTML file, a CSS file, and a JavaScript file.

To work on the opened websites and inject code into them, a content script is needed which is loaded into the website like any other script. We use the content script to inspect the Document Object Model (DOM) of the website. The DOM presents the content of a website as a tree-like structure. The root is the document itself with the <html>-tag, which has the <head>-tag and <body>-tag as children. The children of the <body>-tag are the building blocks of the website. We inspect the DOM to extract the scripts embedded into the website. The source attribute of the script tag is important for our implementation to show the user the origin of the scripts.

The content script cannot communicate with the pop-up script directly as the pop-up script is only active when the pop-up is opened. Therefore, a background script is needed, which runs in the background and can exchange messages with the content script. We use it to extract the origins of the loaded scripts from the content script and store them in a private variable. The pop-up script can access the background script and its variables. This way, the pop-up window can access the origins of the scripts, which are stored in the background script. After accessing the information, we can display it in the pop-up window. The prototype was tested by visiting different websites and evaluating whether all scripts were detected and showed up in the list of the extension.

9.5.3 Final Browser Extension

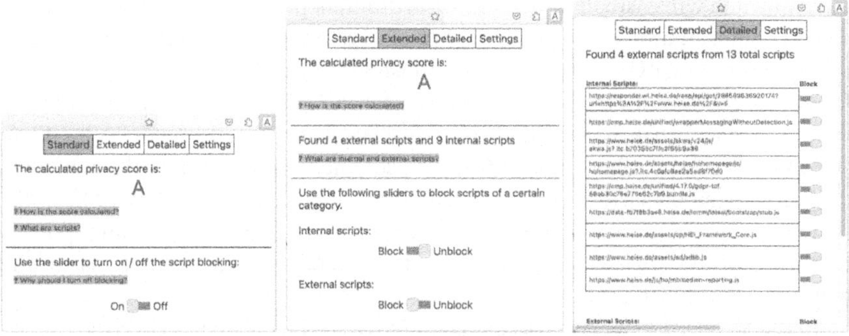

Fig. 9.1 The extension's standard (left), extended (center), and detailed view (right)

After a successful implementation of our prototype, we implemented the features developed during the workshops. We created three different views for the three user groups with different privacy competence as developed during the workshops (cf. Sect. 9.4), which are accessible by using a tab-like menu bar. The views are labeled "Standard", "Extended", and "Detailed". According to workshop 3, the found scripts on a website are classified into two categories, namely internal and external scripts. Internal scripts are scripts that are loaded from the visited website or subdomains thereof. External scripts are scripts from all other sources, so-called third-party scripts. For the application and in the following, we use the term external scripts, since it is more comprehensible for lay users. This classification is a simplification of tracking technologies in general and therefore a limitation that might lead to a less complex user interface. Nevertheless, our design recommendations (cf. Sect. 9.7.3) can still be implemented and evaluated. A more detailed discussion is presented in the limitations (cf. Sect. 9.7.4). In our extension, we disregard the specific function of scripts, which may lead to a wrong classification of websites. While other approaches may classify scripts more accurately, this would not influence the UI elements that we want to evaluate to check whether the support personas like different elements and want different information. Nevertheless, external scripts are a good approximation to find tracking scripts since many external scripts are used for tracking or advertisement. The functionality is often not influenced by third-party scripts which the participants also mentioned during the evaluation.

The privacy score is part of the first two views and ranges from "A" for websites with less than or equal to five external scripts, to "B" for websites with an amount of external scripts between six and 15, to "C" for websites with over 15 external scripts. The thresholds for the different scores were chosen by evaluating the numbers of scripts of different websites. Therefore, it was guaranteed that the participants of the evaluation will easily find websites with different privacy scores. This allows for a good demonstration of the functionality of our demonstrator and is not meant to be a real threat analysis. The scores are illustrated by using different colors: green for the score "A", orange for "B", and red for "C". The icon in the menu bar of the browser also changes with the visited website and shows the letter "A", "B", or "C" in the corresponding color. When the user clicks on the icon in the menu bar, the pop-up window opens. The presented view is the view set as default. This is the standard view for users with low privacy competence. Users with medium privacy competence are shown the extended view. The detailed view is shown to users with high privacy competence.

In the standard view (Fig. 9.1, left), the user sees the calculated privacy score. Below, the user is shown help buttons which provide explanations about the privacy score and the concept of scripts. At the bottom, a slider can be used to turn off the extension for the current website. This might be useful if a website was not loading due to blocked scripts. The extended view (Fig. 9.1, center) also shows the privacy score on top. Below, the number of found external and internal scripts is shown. A help button explains the concept of external and internal scripts. At the bottom of the pop-up, two sliders can be used to block or unblock all external or internal scripts. Finally, the detailed view (Fig. 9.1, right) shows the numbers of found internal and external scripts on top. Below, the user finds a list of all scripts sorted by internal and external scripts. In this view, the user can block and unblock scripts individually. In the settings tab, the user sets the default view.

9.6 Evaluation

Since the goal of this work is to develop the concept of support personas and demonstrate their usage in a development process, the main purpose of the evaluation was to examine whether the application provides good support and user experience to all groups. This would mean that the usage of support personas was successful.

9.6.1 Methodology of the Evaluation

9.6.1.1 Participants

We conducted the evaluation with 31 participants (cf. Table 9.3), ten of whom had also participated in the workshop. This allowed us to evaluate whether the developed application met the expectations of the workshop participants. The new users were chosen to ensure that the extension is usable without prior knowledge from the workshops. Some of the interviews were held in person and some remote (cf. Table 9.3). All interviews were audio recorded to evaluate the Think-Aloud-Method more precisely. The audio recordings were transcribed and analyzed by two researchers to identify further insights about the user experience. It was inspected how the UI was described and what aspects of the UI was liked or not and what the reasons were (cf. Sect. 9.6.2.2). Before conducting the evaluation, we obtained an IRB approval (EK 16/2022) by the ethics committee at our university. In total, we interviewed 19 men, eleven women and one diverse person (average duration of 35 minutes) from Germany. For the statistical Mann-Whitney-U-Test, this means that we can only determine big effects and thus big differences between the groups with a statistical power of 0.72 for an effect size of 0.9 and an error probability of 0.05. Participants were first recruited from the university context (n = 20) and received no financial compensation. These interviews were held between 09/03/2022 and 09/10/2022. To increase the heterogeneity of the sample, we then recruited individuals via the crowdworking platform *Prolific*[4] (n = 7) and the personal and professional networks (n = 4). The participants from *Prolific* received €8 in financial compensation. Prolific is a platform designed to provide samples for scientific studies. Several studies have confirmed the reliability of Prolific and its ability to collect high quality and diverse data [198, 199]. The follow-up study was held between 05/15/2023 and 05/24/2023. Since we reached saturation, i.e., no new findings, and the fact that a sample size of 31 is relatively high for qualitative studies in the research area of human-computer-interaction [200], the sample size of 31 participants was seen as sufficient.

[4] https://prolific.co/

Table 9.3 Overview of the participants of the evaluation

Person	Gender	Age	Field of Study / Job	Workshop	Remote
W1	m	20–24	Electrical Engineering B.Sc.	y	n
W2	m	20–24	Business Informatics M.Sc.	y	y
W3	m	25–29	Environmental Engineer	n	n
W4	w	20–24	Finances M.Sc.	y	n
W5	m	25–29	Computer Science M.Sc.	y	n
W6	m	25–29	Int. Business Studies B.Sc.	n	y
W7	m	20–24	Computer Science B.Sc.	y	n
W8	w	20–24	Int. Business Studies M.Sc.	n	n
W9	m	20–24	Comput. Engineering M.Sc.	y	n
W10	w	20–24	Teaching Student (German, Art)	n	y
W11	m	20–24	Media Informatics B.Sc.	n	y
W12	w	20–24	Computer Science M.Sc.	n	y
W13	m	20–24	Project Manager	n	y
W14	m	20–24	Computer Science M.Sc.	y	n
W15	w	20–24	Computer Science M.Sc.	y	n
W16	m	20–24	Computer Science M.Sc.	y	n
W17	m	20–24	Catholic Theology	n	y
W18	w	20–24	Controller	n	y
W19	m	20–24	Psychology B.Sc.	y	n
W20	w	55–59	Secretary	n	n
W21	m	40–44	Self-employed	n	y
W22	d	40–44	Alternative Practitioner	n	y
W23	m	30–34	Team Leader Logistics	n	y
W24	m	40–44	Commercial Employee	n	y
W25	w	30–34	Self-employed	n	y
W26	m	55–59	Banker	n	y
W27	w	55–59	Homemaker	n	y
W28	w	50–54	Commercial Employee	n	y
W29	w	50–54	Judiciary Employee	n	y
W30	m	45–49	Gardener	n	y
W31	m	35–39	IT System Administrator	n	y

9.6.1.2 Procedure

An important part of the evaluation was the user experience questionnaire (UEQ) [201], which inspects the experience the user had when using the application. All participants were informed about the data collected and processed during the evaluation and signed an informed consent. Before the participants tested the application, we asked for some demographic information (age, gender, study field / job) via a questionnaire form. The participants also answered the OPLIS and the IUIPC questionnaire introduced in Sect. 9.3 to determine their respective support personas. The participants were also asked to answer three questions about their previous experience with script tracking.

While using the application, the participants' voice was recorded and they were asked to think out loud (Think-Aloud-Method [202]). Thus, we were able to receive an immediate reaction to our UI. This data was evaluated to find out what was unclear, where questions arose, or what was surprising to the user. As an introduction, we explained how the pop-up window of the application can be accessed. After this, the participant was asked to visit a few websites and explore the browser extension. After testing the extension, the participant answered the UEQ questionnaire to rate the user experience and answer a few questions about what they liked or disliked. Of special interest was the question whether the user learned something by using our extension. This would indicate that we not only developed a usable but also educational application.

Table 9.4 Results from the Mann-Whitney-U-Test on the UEQ answers

Item	Test Statistics U	z-Value	p-Value
Attractiveness	97	−0.338	0.735
Perspicuity	69	−1.521	0.128
Novelty	80.5	−1.035	0.301
Stimulation	87.5	−0.74	0.459
Dependability	49	−2.367	0.018
Efficiency	110	0.211	0.833

9.6.2 Results of the Evaluation

9.6.2.1 Classification and Preferences of Users

According to the OPLIS questionnaire and our ranking based on the norm table of the entire population, 17 participants were classified as having high, ten participants as having medium privacy competence, and four participants as having low privacy competence (Table 9.5). Thus, we evaluated the application with users corresponding to every identified support persona. We found that 16 of the 31 participants preferred the view recommended to them, while most (22) of the users preferred the "Extended" view due to the combination of simplicity and information. The eleven participants with high competence who preferred the "Extended" view indicated that the "Detailed" view did not provide the score and the possibility to block whole categories of scripts. This shows that the score and therefore, the classification of websites is a feature many users want and also was the feature which was mainly used while trying the application. Nevertheless, six participants preferred the "Detailed" view because they liked the display of the addresses and the possibility to block individual scripts. With one exception, the "Detailed" view was preferred only by people with a high privacy competence, which supports our workshops' result that this user group prefers to receive more information. They also acknowledged the "Detailed" view as a possibility to get further information or more adjustment options. This corresponds with the preference of the developed support persona that people with a high privacy competence like to have the possibility to adjust settings and like to have as many information as possible. In contrast to the workshop results, the evaluation shows that even though users with a high privacy competence like the settings of the "Detailed" view, in the everyday life, they (twelve out of 17) prefer a more simpler view like the "Extended" view. One participant even prefers the "Simple" view in the everyday life. Half of the four participants with a low privacy competence preferred the "Standard" view while the other half preferred the "Extended" view after reading the explanations in the "Standard" view and familiarizing with the extension. Regarding the educational aspect of our browser extension, all 21 participants without prior knowledge about scripts and script tracking stated that they learned what scripts are. Additionally, they stated that they understood the difference between external and internal scripts. This mirrors the considerations from the workshops and our literature findings that lay users with a low privacy competence need easy-to-understand explanations. In addition, the fact that users who did not know about script tracking read the explanations in the "Standard" view shows the functioning of our UI design. The explanations allowed the users to understand and use the "Extended" view. In addition, the extended view provided the settings which nine of the ten participants with medium privacy competence wanted.

Table 9.5 Overview of the results of the questionnaires

Person	Privacy Competence	Privacy Concern	Knew About Scripts & Script Tracking Before	Preferred View
W1	high	high	yes	Detailed
W2	high	high	yes	Detailed
W3	medium	high	no	Extended
W4	high	high	no	Extended
W5	high	high	yes	Extended
W6	high	high	no	Detailed
W7	medium	high	yes	Extended
W8	high	high	no	Extended
W9	high	high	yes	Detailed
W10	medium	high	no	Extended
W11	medium	high	no	Extended
W12	high	high	yes	Extended
W13	medium	high	no	Detailed
W14	high	high	no	Extended
W15	high	high	yes	Extended
W16	high	high	yes	Extended
W17	high	high	no	Extended
W18	high	high	no	Detailed
W19	medium	high	yes	Extended
W20	medium	high	no	Extended
W21	medium	high	no	Extended
W22	high	high	no	Standard
W23	medium	high	no	Extended
W24	high	high	no	Extended
W25	low	high	no	Standard
W26	high	high	no	Extended
W27	low	high	no	Standard
W28	medium	high	no	Extended
W29	low	high	no	Extended
W30	low	high	no	Extended
W31	high	high	yes	Extended

9.6.2.2 User Experience

Since the objective of our work was to develop an application that supports the user best and is thus easy to use, the results of the UEQ questionnaire were of particular interest. It measures six different aspects: attractiveness, perspicuity, novelty, stimulation, dependability, and efficiency.

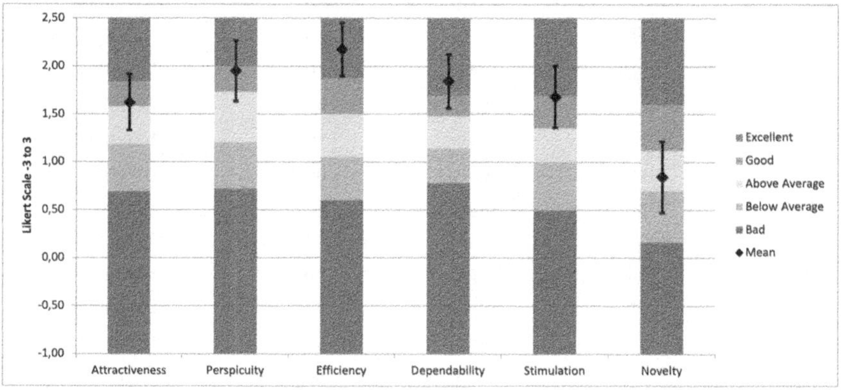

Fig. 9.2 Results of the UEQ-questionnaire

The results (cf. Fig. 9.2) show that our browser extension has good to excellent scores in all aspects except for the novelty rating. The lower score for the novelty is not surprising since our application combines features from existing browser extensions to present the appropriate information to the user. The perspicuity score shows that our application is easy to use, to learn, and comprehensible. This is supported by the statements of the participants who found the extension "very intuitive" (W5, W7, W10, W15, W16, W21, W24), "easy to handle" (W29) and the explanations "easy to understand" (W2, W3, W6, W8, W10, W15). Many participants also liked the given privacy score in the menu bar which gives good information to see if a website is more privacy-protecting or not (W2, W4, W9, W11, W18, W20, W26, W31). A point of criticism was the missing score range for the privacy score (W7, W9, W31). Considering the feedback we received, attractiveness and stimulation could be improved with the implementation of a more modern design and the use of toggle switches instead of slide switches.

We performed a Mann-Whitney-U-Test [203] to evaluate whether the user experience varies between workshop participants and others. The results can be seen in Table 9.4. In the Electronic Supplementary Material (A.2.4), the detailed results can be seen. We found that for the two-sided significance level of 0.05, there is only a difference between both groups regarding the dependability. The participants who also take part in the workshops assessed the dependability higher, which shows that they felt more comfortable when working with the application. The other aspects of the user experience showed no differences between both groups. In conclusion, 26 out of the 31 participants stated they would use the browser extension in their everyday life. Most of them concluded that the tool would be easy to use and conveys a good impression of the potential tracking that is performed on a website. Although most of them would not use it to actively block scripts, the first impression of the given privacy of a website shown by the privacy score would be reason to use the browser extension. One participant (W1) told that he already has a similar application installed which has more functions and thus, would not use the application. Another participant (W16) stated that he would consider to change to our extension due to the simplicity and clarity. W21 and W31 stated that they would use it if they saw an advantage to other applications or the built-in tracking protection of the browser.

9.7 Discussion

The core contribution of our work lies in the development of the concept of support personas that are used to classify users in different categories to provide them with the support they need and want when using PETs. We verified the concept by applying it to a browser extension that informs the user about scripts and script tracking during web browsing. The evaluation results show that the extension is not only easy to use but also educates the user.

9.7.1 Methodological Contribution

In our research question, we asked how users of PETs with different backgrounds can be supported through UI components. The related work shows that there is a demand for applications that support the individual user in making more informed decisions [172]. Additionally, it was shown that in applications providing explanations for less experienced users, they often struggle to understand the terms used in the explanations [45, 46]. This underlines that more research is required to provide support for different types of users. We concluded that in order to provide customized support to each user, we would need to divide users into distinct groups. Since the concept of personas has proven to be helpful in requirements engineering [40], we developed support personas that represent user groups based on the support a user wants and needs from an application. These are the main contribution of the paper. We based the support personas on the user's privacy competence and concern. The needs of the different personas were then developed and refined in workshops for participatory design.

Support personas have some common aspects with privacy personas developed in previous works. The privacy personas were either based on users' privacy concern or privacy competence, together with users' privacy behavior [160, 170]. The difference between privacy personas and support personas lies in the purpose for which they were developed. While privacy personas were developed to classify users with regard to their attitude towards privacy, the protection of their data, and the behavior when using technology, support personas aim to elaborate how different user groups can be supported by providing them with the best user experience when using PETs. Additionally, support personas allow for UI customizations adjusted to the users' needs and not only their current privacy behavior. The definition of a persona includes not only the items they are based on, but also the motivation, attitude and goals of a user. The motivation and goals of support personas are differing from those of privacy personas because of the different focus. The demonstrator PRIVACYASSIST shows that we could address the different user groups by using support personas in the development process. Other developers of PETs can also use the developed support personas and the UI guidelines (cf. Table 9.2) to develop PETs which are more usable for all different users. Especially by considering the support personas, the developers will have the needs of all the different users in mind. This will help to develop applications not only for a specific target group but for all users. That will lead to PETs, which are not only usable by security experts but also by lay users. Thus, the support personas are an important contribution to the development of a UI that best supports the user while appropriating new PETs and the key contribution of our paper.

9.7.2 Support Personas During Development

To evaluate the concept of support personas in a development process, we designed a browser extension, PRIVACYASSIST, that provides users with information about script tracking. It identifies third-party scripts that are often used for script tracking and, therefore, informs users about their usage on visited websites (cf. Sect. 9.5.3). The necessity of such an application was derived from related work. It was found that internet users often only have basic knowledge about online tracking [181] and that a UI providing the user with more information about the security of a website is needed [182]. It was also shown that a privacy dashboard presenting details about potential privacy threats improves users' privacy perception [185].

To involve different stakeholders in the design process [197], the design of the extension's UI was conceptualized during three workshop sessions with a total of 15 participants. The contribution of the workshops were guidelines that should be adhered to when developing a usable application. Thus, we complement the design guidelines developed by existing works [159, 162, 177, 178]. Additionally, we found that the UI should be primarily based on the users privacy competence to provide the best support possible. In comparison to existing approaches, PRIVACYASSIST combines the functionality of them and provides different user interfaces for different support personas. During the evaluation, we verified that our application is easy to use and educational for the different support personas. We showed that different users prefer different UIs that present the information they want and need. Users with less knowledge about scripts considered the explanations we provided in the application as very helpful. More advanced users liked the detailed information and configuration options provided in the "Detailed" view. This shows that support personas are a valuable contribution to the development process of PETs.

9.7.3 Recommendations

The related literature [162, 172, 177] suggests that users' understanding of security issues and awareness of their security needs are critical when designing user interfaces. Therefore, we designed our support personas according to these two factors, namely privacy concern and privacy competence. Future work can use the support personas to examine user groups in different privacy scenarios. Every support persona requires different features of a privacy enhancing user interface. These should be addressed by developers of PETs. Regarding privacy concerns, users with low values may become frustrated if the user interface hinders their workflow, such as by blocking suspicious scripts that remove essential parts of a website. Conversely, users who highly value privacy would likely tolerate a certain degree of performance drawbacks, as long as their privacy is not compromised. Therefore, the default values in PETs should be set according to the users' level of privacy concern. When it comes to the privacy competence of users, there should be different UIs for each user group, e.g., providing easy to understand explanations for novices, which ideally increases their privacy competence, and extensive information and settings for experts. However, users should not be restricted to a single UI and have the option to switch views if they desire. For instance, while an expert user may occasionally desire detailed information, it may be advantageous to provide a simpler view for the majority of the time. To account for individual preferences, we suggest involving users early in the development process, allowing them to provide feedback and contribute to design decisions, to avoid misinterpretations early on and increase the UI's usability. Easy to understand explanations also increase the privacy literacy of the users which helps users to secure their privacy. For users, we recommend that they should inform themselves about privacy topics to be more aware and knowledgeable and make more secure decisions. For browser tracking, users should be aware that data about them or about their browsing behavior is often distributed to third parties. Our developed application can help assess whether a website runs scripts that can possibly track users. Based on this information, users should decide which information they provide to the website or whether they want to visit another website. Especially personalized information can help users to get the information they need to make more secure decisions.

9.7.4 Limitations

While the participants in the workshops had different study background, they were students from the researchers' personal and professional networks, which could result in different biases. Although the presentation of existing applications helped the participants to get an idea of the application to be developed, all participants only used existing modules while completely new ideas were not developed. This limited the novelty of the extension. Another limitation lies in the privacy score we calculated for the visited websites. Currently, the score's calculation is based on the number of found third-party scripts. Although those scripts are often used for tracking the user, they can also be used to load content, e.g., new fonts, which is not relevant for security. Another problem is that websites like *www.youtube.com* do not use third-party scripts but track the user using their own scripts. For both cases, it would be beneficial to know the exact purpose of a loaded script. This feature was also requested during the evaluation, as it provides knowledge crucial for deciding whether to block a script or not.

The distinction between first- and third-party scripts represent a simplification of a complex tracking ecosystem (e.g., including browser fingerprinting, cookies, or tracking requests) to a certain degree. However, it enabled to develop a proof of concept of how support personas can be used in the development process of PETs. With a more fine-grained distinction, e.g., in essential, functional, and marketing scripts, more visualizations and settings would be necessary. Additionally, further explanations would be needed which could make the extension more complicated. All this must be evaluated in practice, although we think that with our given recommendations like the usage of easy language, the development of an extension for all users is still possible. Additionally, PRIVACYASSIST can be extended by implementing more features, like the visualization of additional tracking elements.

Finally, another limitation is that all participants were ranked as having high privacy concern. One reason could be the used questionnaire. While a good construct validity and reliability was shown [204], the questionnaire could show weaknesses in practice. Eventually, a more specific questionnaire to the topic of script tracking could have solved the problem, but this would have restricted the generalizability of the support personas. Another reason could be that users, when being asked, rate their privacy concern higher than it actually is. This can be in relation to the privacy paradox.

9.7.5 Future Work

Future work could explore the concept of support personas further and draw on it to build more secure applications that can be utilized by users with different knowledge and background. To further refine the concept of support personas, another evaluation classifying users not only by their support persona but also by their privacy persona, e.g., using the questionnaire by Biselli, Steinbrink, Herbert, Schmidbauer-Wolf, and Reuter [171], may be conducted to compare both concepts. Additionally, PRIVACYASSIST could be extended to support more security and privacy related concepts of web browsing, e.g., cookies. A modular and dynamic application could also help to make the application more personalized. During our evaluation, some users mentioned that it would be helpful to see the purpose of an individual script in the detailed view. Future research could try to analyze scripts and explain their function to the common user. This could enhance the accuracy of the privacy score by building it on the purpose of the identified scripts. For further feedback, a feedback option for end users can be implemented. An implementation of a domain-to-entity mapping could improve the helpfulness of the source information of scripts.

9.8 Conclusion

More secure critical use cases of computers in peoples' everyday life require secure applications. A problem of such applications is that they are not easy to use for lay users. One reason being that the applications are often developed with advanced users as primary target group. Thus, users with limited knowledge receive less support and, e.g., do not understand provided explanations. So, we developed a strategy to support every user of secure applications. The core of this strategy is the concept of support personas that classify users by the support they need. To demonstrate how this concept can be implemented in a development process, we developed PrivacyAssist, a browser extension that supports the user in the context of found scripts in a participatory design approach. In three workshops, tailored UIs for three different user groups were created. During the evaluation, we found that offering different UIs for different user groups is very helpful and provides every user with the level of support needed. We also verified that our application is easy to use and educational. Thus, we conclude that support personas are a suitable strategy to support users with different backgrounds when using PETs.

Acknowledgments Funded by the Deutsche Forschungsgemeinschaft (DFG, German Research Foundation)—SFB 1119 (CROSSING)—236615297. This research work has been funded by the German Federal Ministry of Education and Research and the Hessian Ministry of Higher Education, Research, Science and the Arts within their joint support of the National Research Center for Applied Cybersecurity ATHENE.

Paper E: Building Trust in Remote Attestation through Transparency - A Qualitative User Study on Observable Attestation

10

Abstract

Internet of Things (IoT) devices have become increasingly important within the smart home domain, making the security of the devices a critical aspect. The majority of *IoT* devices are black-box systems running closed and pre-installed firmware. This raises concerns about the trustworthiness of these devices, especially considering that some of them are shipped with a microphone or a camera. Remote attestation aims at validating the trustworthiness of these devices by verifying the integrity of the software. However, users cannot validate whether the attestation has actually taken place and has not been manipulated by an attacker, raising the need for HCI research on trust and understandability. We conducted a qualitative study with 35 participants, investigating trust in the attestation process and whether this trust can be improved by additional explanations in the application. We developed an application that allows users to attest a smart speaker using their smartphone over an audio channel to identify the attested device and observe the attestation process. In order to observe the differences between the

Original Publication *S. Linsner, K. Demuth, S. Surminski, L. Davi, and C. Reuter, "Building Trust in Remote Attestation Through Transparency—A Qualitative User Study on Observable Attestation," Behaviour & Information Technology (BIT), pp. 1–21, 2024, Publisher: Taylor & Francis*, ISSN: *0144–929X.* DOI: *10.1080/0144929X.2024.2374889. [Online]. Available:* https://doi.org/10.1080/0144929X.2024.2374889

Supplementary Information The online version contains supplementary material available at https://doi.org/doi.org/10.1007/978-3-658-47445-4_10.

S. Linsner, *Privacy Preserving Data Management*, Technology, Peace and Security I Technologie, Frieden und Sicherheit,
https://doi.org/10.1007/978-3-658-47445-4_10

applications with and without explanations, we performed A/B testing. We discovered that trust increases when additional explanations of the technical process are provided, improving the understanding of the attestation process.

10.1 Introduction

Due to their variety, scalability, and simplicity, Internet of Things (IoT) devices face many security challenges that are difficult to cope with [205–207]. For instance, the limited computing power of IoT systems impedes the execution of antivirus software [206]. Furthermore, the majority of IoT devices are equipped with closed and pre-installed firmware, which makes them black-box systems. As a consequence, the user has little assurance on whether the system is indeed operating the intended task and not performing malicious tasks (e.g., spying). Incidents such as the well-known Mirai botnet, which targeted consumer devices like routers and IP cameras, have demonstrated the large impact of attacks against vulnerable IoT devices [208]. Hence, it comes as no surprise that users have little trust in IoT devices, especially if these devices are equipped with microphones or cameras like smart speakers or smart screens [209, 210].

One prominent research direction to increase trust and security of IoT devices is *remote attestation* [211–216]. Remote attestation is a security service that allows a device, the verifier, to verify the integrity of a remote, untrusted device, the prover. To perform the attestation, the prover inherits a self-measurement routine. Remote attestation is typically implemented using a challenge-response protocol: the verifier sends an attestation request to the prover, which performs a self-measurement and sends the result back to the verifier [217]. Remote attestation can assure different security guarantees, varying from firmware integrity to more complex run-time properties such as control-flow [218] or data-flow integrity [213, 219, 220], thereby detecting attacks like persistent malware [217], return-oriented programming attacks [221], and even data-oriented programming techniques that do not alter the control flow [222]. Remote attestation is an abstract concept. Its security relies on cryptographic protocols and functions, to which users are often oblivious. Existing research shows that users tend to discard or circumvent mechanisms they do not understand or trust [223]. Users, however, are a critical part of the IoT ecosystem as they are responsible for the devices, and their integrity. Many security mechanisms eventually rely on the user correctly using them, such as virus scanners, firewalls, and passwords [224]. Today, many security risks in IoT devices already arise due to incautious users that utilize insecure or default passwords, do not apply updates, or configure devices insecurely [225].

However, so far, all remote attestation schemes have been black boxes, which can lead to misuse by users who do not understand and trust the system (backfire-effect [226]). Bringing the user into the context of remote attestation can increase trust in these security mechanisms. Therefore, HCI research is needed to design new usable security solutions. Recently, the first user-observable remote attestation scheme was presented [227]. The authors conducted a user study claiming good usability. However, this still raises the question of whether remote attestation can help to increase trust in IoT devices, e.g., by removing security concerns like the leakage of sensitive information to third parties [228, 229]. In particular, it has not been investigated how users treat compromised devices and devices that were formerly compromised but have been restored to a benign state. Hence, it remains largely unclear whether users trust the abstract cryptographic concepts of remote attestation schemes, raising the demand for usable security mechanisms.

In this paper, we investigate for the first time whether user-friendly and understandable techniques as well as transparency enhance users' understanding and trust in remote attestation. Examples of user-friendly and understandable techniques are explanations that are not only understood by experts and an easy-to-use design of the application. The concept of trust used in this study entails the users' trust in the correctness and effectiveness of the attestation process, as well as the user's trust, that the attested device runs as intended and no third-party privacy violations are possible. We aim to use transparency as an HCI intervention mechanism to inform the users about the attestation to enable the users to comprehend the process and the promises of remote attestation. Therefore, our main research question is: **'Does transparency enhance trust in remote attestation?'** We implemented an application of a user-observable remote attestation scheme that has recently been developed [227] into a custom-made demonstrator including a real smart speaker and evaluated it in a user study.

10.2 Background and Related Work

In this section, we provide background information on the topics of attestation and trust in IoT devices and present related research. We conclude by pointing out the research gap resulting from the presented related work.

10.2.1 Remote Attestation

Remote attestation allows the verification of the integrity of an untrusted device [230]. For example, a server can verify whether a client system has an uncompromised version of a software. The main problem in attestation is obtaining a trustworthy measurement of an untrusted device. Even if the system is completely compromised, this measurement must not be altered. To this end, the attested device requires logic to prove its integrity, which is why it is referred to as *prover*, while the device verifying the reports of the prover is called *verifier* [230]. There exist three categories of attestation schemes, depending on their architecture, that is, hardware-based, software-based, and hybrid attestation schemes [231]. Hardware-based attestation schemes rely on a trusted computing base inside the untrusted device [232–235]. Software-based attestation schemes rely on precise timing measurements to perform the attestation [236, 237]. Hybrid attestation schemes combine hardware and software designs to perform trusted self-measurements [217, 238, 239]. These attestation protocol schemes were developed for a setting in which a trusted verifier attests one or more prover devices. Typically, remote attestation schemes inherit a challenge-response protocol. The verifier starts the attestation by sending a nonce to the prover to ensure freshness and prevent replay attacks [217].

For IoT devices, there exist many attestation schemes with different advantages. Since IoT systems often consist of many different devices, most existing remote attestation schemes lack the scalability to attest a network of IoT devices. Therefore, the Collective Remote Attestation (CRA) scheme was developed which can perform the attestation of IoT networks [240]. Kuang, Fu, Yu, Yang, Su, and Zhang [211] propose a *many-to-one* remote attestation scheme that eliminates the need for a fixed verifier and thus, reduces the possibility for an attacker to tamper with the attestation process. For the specific threat of roving malware, i.e., malware that relocates itself in the memory, Aman, Basheer, Dash, Wong, Xu, Lim, and Sikdar [230] developed an attestation scheme that can detect this malware. During most remote attestation schemes, the device must not be interrupted. This means that in order to attest a network of devices, all devices must interrupt their normal operation until the attestation is performed. In order to avoid such a network-wide interruption, Dushku, Rabbani, Conti, Mancini, and Ranise [212] developed an asynchronous attestation scheme allowing parts of the network to operate normally while another part of the network is being attested. Attestation is not only used for IoT devices but inter alia also used for hardware security tokens (HST). It was shown by Pfeffer, Mai, Dabrowski, Gusenbauer, Schindler, Weippl, Franz, and Krombholz [241] that existing authenticity checks have usability issues that cause users to base their trust on ineffective security features such as visual damage of a product. They suggested to use combinations of automation and user interaction to improve the security of attestation procedures.

10.2.2 Trust in IoT

Trust in IoT devices is essential since they gain more importance in the everyday lives of many people. Smart light bulbs, for example, create easily adjustable ambient light, smart speakers that can be controlled by voice make it easy to adjust the music playing in the background and smart thermostats allow to control the temperature at home from anyplace at anytime. However, IoT devices are notoriously insecure [207]. Manifold prominent examples, such as insecure security cameras [242], attacks on smart speakers [243], and large-scale attacks such as the Mirai botnet [208] have customers worried about the security of home IoT devices [244–246]. Users are also worried that sensitive or personal information could be leaked to third parties [228, 229, 247, 248]. Furthermore, IoT devices are designed as black box systems to the user [210] and are used *as is*, which is why users do not have control over the software running on the devices. Research showed that security, besides other factors, is an important aspect for people in order to trust an IoT device [249]. The generally positive attitude of the user to use an IoT device [250] can then be affected by this trust [251] as well as the complexity associated with the usage [252]. Recent work has also shown that many users are unaware of the privacy risk posed by devices that do not have a camera or microphone [209] as well as IoT devices in general [253–256] leading to incomplete threat models [257, 258]. It was also shown that users often express low privacy concerns [246, 259] even when using applications which pose security and privacy threats [247, 260]. This is amplified by the fact that many owners of smart home devices do not fully understand their own data practices [253, 259, 261]. Consequently, while users see privacy and security risks with devices that have built-in cameras or microphones, they often trade them for functionality [210, 254, 262] or availability [248, 259]. This shows the need for solutions that increase the privacy and security of users but minimize the costs of using them [262, 263]. Especially, it was shown that these solutions need to establish not only privacy and security but also trust in the devices [264].

10.2.3 Research Gap

Trust in an IoT device plays an important role during the adoption process [249, 265]. Since security is an integral aspect for people in order to trust an IoT device, mechanisms which increase the security of IoT devices are needed. A key part of security responsibility lies within user awareness as careless behavior can lead to security breaches [266]. Consequently, HCI research towards aiding user security awareness and behavior in dealing with IoT is necessary. While there are mechanisms enhancing the security of consumer IoT devices [211, 212, 230, 240], there is also a need for mechanisms that are easy for end users of the devices to use [244]. Here, one way to improve the security of IoT devices is remote attestation. A problem with many traditional remote attestation concepts is that the remote attestation process is completely nontransparent to the user. Thus, the user is neither part of the attestation process, nor can observe the attestation process, and thus often has an incomplete threat and security model [210, 253]. Previous work showed that it is possible to make remote attestation user-observable [227]. However, it has yet to be proven, that this increases the trust in this process. Therefore, we investigate whether users trust the attestation process and explore whether making it more transparent can further improve the trust in remote attestation. Furthermore, we examine another important aspect of remote attestation, concerning the issue of how to deal with a compromised device. Remote attestation by design can only detect a compromise but does not specify how to react to it. In end-user scenarios, it is therefore relevant to determine whether users trust the attestation and whether they trust benign—but formerly compromised—devices.

10.3 Attestation Application

In this section, we present the attestation application used for our study. Based on a remote attestation scheme [227], we built an application that can be used to simulate the attestation process of a smart speaker. The reason for choosing a smart speaker is that these devices have a widespread use and have microphones installed, which poses risks for privacy but enables the attestation via an audio side channel.

10.3.1 User-observable Attestation

SCAtt-man is a remote attestation solution specifically designed for IoT devices [227]. To solve the problem of lack of device authentication in software-based attestation and ensure reliable one-hop communication, side channels such as sound and light are used. In SCAtt-man, the missing device authentication is replaced by the user. By using user-observable communication channels, the user can monitor the attestation process and identify the device currently being attested. Figure 10.1 shows a depiction of this process. To perform the attestation, users can use their smartphones with an attestation app. For practical usage, this functionality could potentially also be integrated into the configuration apps already used for smart speakers, such as the Amazon Alexa app. The app guides the user through the attestation process. First, the user presses the attestation button on the device they want to attest. Then, the app performs the attestation and asks the user to wait. After completing the attestation, the app informs the user about the result. If the attestation fails, the app shows the user how to resolve the problem. A failed attestation can have different reasons: There has been a communication failure, either on the audio communication or the network transmission, or the device has actually been compromised. To rule out transmission problems, the user can repeat the attestation process. After multiple failures, the user can resolve a compromise by resetting the device to factory settings and repeating the attestation process to verify the device's integrity. In an initial user study, it was shown that SCAtt-man has good usability and people trust the attestation [227]. However, the prototype only had limited smart speaker functionality as the focus was on the development of the attestation scheme.

Fig. 10.1 The attested device communicates via audio with the smartphone. The user observes this process and can identify the attested device

10.3.2 Smart Speaker for Attestation

To tackle the problem of limited functionality, we re-implemented the SCAtt-man attestation scheme into a state-of-the-art smart speaker. As smart speakers such as Amazon Alexa and Google Home are closed systems, a full implementation of SCAtt-man was not possible as this requires full access to the hardware and software. Therefore, the only solution is to imitate the behavior of the SCAtt-man attestation scheme.

We developed a smart speaker application for the Amazon Alexa voice assistant that resembles the SCAtt-man attestation process. Amazon Alexa is the most popular smart speaker with a market share of 28.3% [267]. Amazon Alexa offers a wide range of services and keywords and can be used for daily tasks, such as setting timers and alarms, creating shopping lists, telling jokes, or playing music. Many IoT devices allow integration so that Amazon Alexa can be used to control smart homes with voice commands. It can also be used for more complex tasks such as online shopping. Amazon Alexa allows third-party apps and integrations. These apps are called skills and can further extend the functionality of Amazon Alexa. Our implementation consists of three parts: (1) An Alexa skill to implement the attestation behavior of the smart speaker; (2) The installation of the attestation app on the smartphone running Android 13; (3) A coordination service to allow communication between the app and the skill.

Figure 10.2 shows the architecture of our setup. For the smart speaker, we used a standard Amazon Echo Dot. We built a smart home setup with a smart lamp and a smart plug. For the attestation, we required a button to start the attestation. For this, we used a Flic button that communicates over Bluetooth LE using a companion app. We used a hardware button to trigger the attestation because we wanted the attestation feature to be easily accessible to the users. Since we investigate user-observable attestation, it would be detrimental to hide the attestation feature inside a menu. A voice command would be an alternative, but in a real-world scenario, this would require every user to read the manual to find this function. We ran the app for the Flic button as well as all other management apps on a dedicated Android smartphone to separate these apps from the user. We designed a custom 3D-printed stand for the Echo Dot and the Flic button to provide the look and feel of a complete, professional device. Figure 10.4 shows a photo of the final setup. Furthermore, we enhanced the case with a reset button, to perform a reset process during the user study. The Alexa app, like any other Alexa app, was deployed to Amazon AWS. It was not possible to implement actual audio communication between smartphone and smart speaker due to limitations in audio processing in the Alexa API. We, therefore, developed a coordination service that allows coordination between the app and the Alexa skill. This setup allows straightforward usage anywhere as long as Internet connectivity is available. We implemented this service using PHP and an SQL database. This service also has a dedicated web interface to allow the experiment supervisor to influence the attestation result (success/failure) and to activate the extended explanations in the smartphone app, thereby allowing simple control of the complete experiment. We modified the smartphone app so that instead of audio communication it uses our coordination server to start the attestation process. Furthermore, it was extended with explanations. These explanations as well as the result of the attestation can be controlled by the coordination service. The setup for our user study is described in detail in Sect. 10.4.1.3.

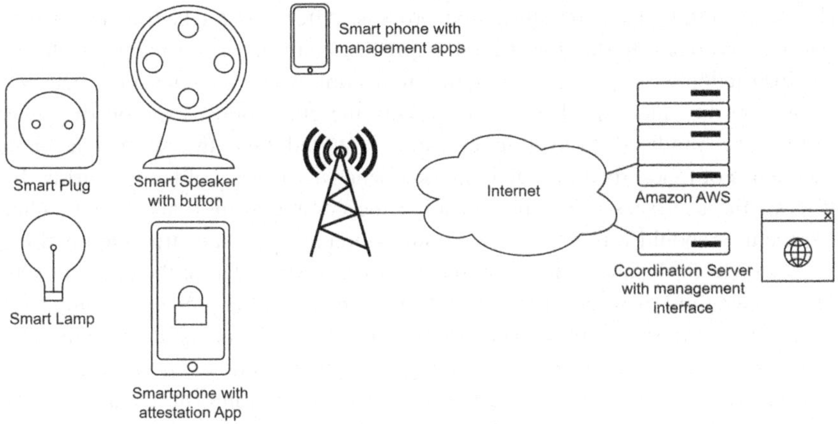

Fig. 10.2 The architecture of our implementation

10.3.3 Application Design

We modified the smartphone application [227] to make it work with our simulated attestation process.

Layout In Fig. 10.3, the different screens of the application for different steps in the attestation process are shown. On the start screen, the user sees the application logo and two buttons. After clicking the explanation button, a pop-up window is shown which explains the attestation process using the following text:

> *"Attestation is used to check the security of devices. To do this, you send a device a task that it must solve and return the result. If the result is correct, the device is safe. However, how can you get a reliable reading from a hacked device? After all, if there is a virus on the device, it can influence the measurement and claim that everything is OK. In order to prevent this, the device is sent a mathematically so complex task that the capacities of the device are fully utilized by it. In this way, any additional load caused by malicious code is noticed immediately. That is why the measurement takes a few seconds. However, it is necessary to prevent the device from forwarding the task to another device, on which the solution for the task could be calculated. Therefore, during the scan, the device is automatically disconnected from the Internet and communication is performed over sound. This way you can tell which device is responding and no virus can forward the task."*

The other button starts the attestation process. After it is clicked, a screen with instructions on how to start the attestation process on the smart speaker is displayed. The instruction is supported by a picture indicating where the attestation button is located on the smart speaker. After clicking the attestation button on the smart speaker, the simulated audio communication is started. First, the smartphone emits a sequence of tones, then the smart speaker sends a sequence of tones. During this process, the user sees a picture indicating the communication and a text saying that the user should wait until the communication and thus the attestation is done. The progress of the attestation is visualized by a progress bar at the bottom of the display (see third picture in Fig. 10.3). When the attestation process is completed, the user is shown the result of the attestation. In case of a successful attestation, the user observes a green result screen saying that the attestation was successful (see fourth picture in Fig. 10.3). The 'failed' screen in red color informs the user that the attestation may have failed either because the device is compromised or because of a transmission error (see fifth picture in Fig. 10.3). In this case, it is recommended to the user to conduct the attestation again. For this purpose, the user can get back to the start screen by clicking the respective button. If the attestation repeatedly fails, the user should reset the smart speaker by pressing the reset button. A button in the

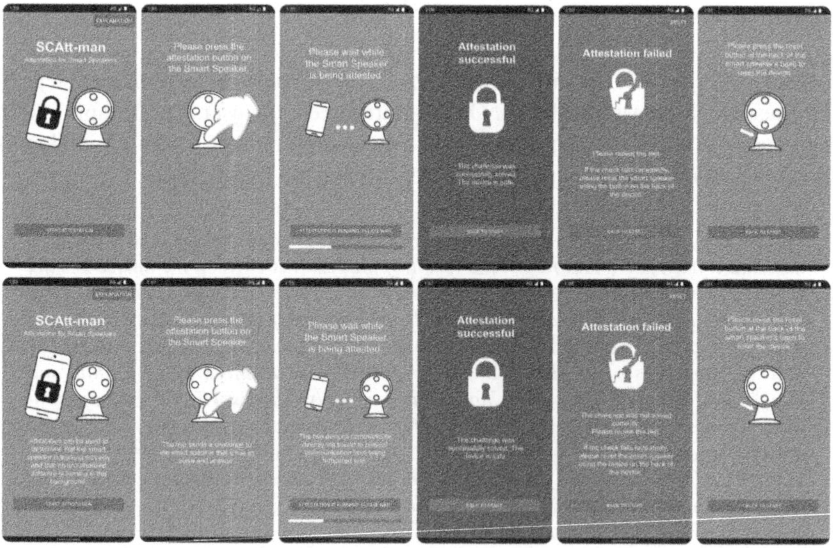

Fig. 10.3 Standard version of the application (top), transparent version (bottom)

top left corner of the 'failed' screen can be pressed to get further information on where the reset button is located (see sixth picture in Fig. 10.3).

Besides the standard version of the application, we also developed a transparent version of the app (cf. Fig. 10.3 bottom row). This application consists of the same screens as the standard version but with additional explanations. At each step, additional explanations are displayed describing what is done in the background to inform the users about the mechanism behind the attestation. Solely the success screen and the screen providing explanations on the location of the reset button were not altered. We developed this version of the application to be able to examine whether additional explanations can improve the understanding of and trust in the attestation process.

10.4 Qualitative Evaluation

In this section, we present the design of our study on user-observable remote attestation and the results from the 35 interviews we conducted.

10.4.1 Methodology

As mentioned in Sect. 10.1, the main research question we aim to address is: **'Does transparency enhance trust in remote attestation?'** In order to answer this question, it is necessary to understand how trust is generated. We investigate trust in two dimensions: The trust users place in the correctness of the attestation and the trust users place in their devices. Literature indicates that users have less trust when confronted with a black-box system or when they do not understand how their data is processed [210, 261]. This can be translated to: Users are more likely to trust concepts that they can comprehend. Therefore, understanding a concept is key to building trust. As related work shows [210], software systems are often black-box systems, which makes it hard to understand how a mechanism is working. We aim to make the internal functionality transparent to end users, in order to support their comprehension of security mechanisms. However, literature [268] shows, that too much information can cause information overload and confuse the user, which would be detrimental to trust. From this, we derive our first sub-question: **'How does transparency affect the users' understanding of remote attestation?'**

In this study, we use two approaches to provide the users with transparency: First, we provide users with two different versions of the app: a version providing just the basic functionality, serving as a control group, and a transparent version which contains information on what is happening during each step. Other researchers showed that too much information can lead to information overflow and thus contradict our aim of educating the users about security mechanisms. Therefore, we chose to add small text in every step of the attestation process, explaining what is happening at the moment. Due to space limitations, we only used short text, which should not be overwhelming for users. The investigation of abstract concepts like understanding and trust is prone to biases such as the acquiescence bias [269]. To combat this, we use a two-fold approach: We construct questionnaire items (see Electronic Supplementary Material (A.3.3.2) to measure the self-reported understanding of the participants and secondly, we interview the participants after the interaction with the demonstrators and ask them to reiterate how they think the attestation works. We then can compare the self-reported understanding with their ability to reiterate the process in the experiment. This way we can limit biases and validate the self-reported understanding to get a more detailed impression of the actual understanding of the participants. Questionnaire items B1 (*I understood how the app works*) and B2 (*I understood how the process behind attestation works*) investigate the degree of understanding of the process of remote attestation. We decided to split these items in two, because participants could confuse the usage of the app with the theory behind the process if not asked for both separately. All items are presented with a 6-point Likert scale and a "no answer"-option. To answer the first sub-question for the first transparency condition, we compare the derived level of understanding of both groups of the A/B testing.

For the second approach, we embedded an explanation into the app (cf. Sect. 10.3.3). On the title screen, users could use the "explanation"-button in the top right corner to read the explanation text on how the remote attestation works. This text describes the process behind the attestation in an easily accessible way and was designed for users who would like to learn more about the process. Similar to the first examination, we investigated the effect of reading the explanation text on the understanding of the attestation process by utilizing the combined approach of questionnaire items and interview to compare the understanding between the participants who read the explanation text and those who did not.

Since we are investigating whether transparency improves understanding, we also have to investigate, whether understanding leads to more trust in order to answer our main research question. From this we derive our second sub-question: **'How does the users' trust change by gaining a deeper understanding of remote attestation?'**

To investigate this, we interviewed the participants about their prior experience with smart speakers and how they perceived the attestation process. For this purpose, we utilized the guiding questions from our interview guideline ('Do you use smart home devices at home?'; 'How does attestation change your interaction with smart home devices?'; 'How well were you able to understand how the attestation process works? (Please reiterate in your own words how you think the attestation works.)'). Depending on their reports on prior knowledge, as well as their ability to reiterate the process, we validated whether key features such as audio transmission were understood and whether this understanding increased the participants' trust in the attestation.

Finally, to test the trust in remote attestation we have to consider the case of a compromised device. Therefore, our third sub-question is: **'Do users trust formerly compromised devices, that have been restored and verified with remote attestation?'**.

To answer this, we use two smart speakers in our setup. The user interacts with both, but the first one that is attested should succeed the attestation, while the second one should fail. Then a reset scenario is conducted and the attestation is repeated. In the interview, we investigate with our guiding questions (see Electronic Supplementary Material A.3.1) whether the users would change their behavior towards the formerly compromised device and which role the attestation process plays in their answer.

10.4.1.1 Ethical Considerations

Since we are gathering personal and personal-relatable information from our participants, we consulted our university's ethics committee and obtained the approval (date of approval and approval number deleted for anonymization). To inform the participants about the aims of the study and the data to be collected, we provided an informed consent form. Following the guidelines of the ethics committee, the form contained information on how the data would be processed, when it would be deleted, and whom to contact to withdraw consent. Personal data collected from the participants were age, gender, education degree, and their IBAN to be able to transfer the compensation of €20. The consent form was mailed to the participants before the evaluation to allow them to read it unaffected by a study situation in which it would be more difficult to reject the terms of the consent form. Before each session, participants were given a hard copy of the consent form, and they were given time to review and sign the document. At the end of the study session, the participants were informed that the attestation process was simulated and no actual attestation was performed. The reason for this is that in our evaluation scenario, one device should be compromised. Since we consider it unethical to use real malware in this evaluation setting, we chose to simulate the attestation results. The participants were informed about this fact at the end of the sessions and asked whether this would change their view on the attestation process. None of the participants stated that it would change their opinion on the attestation process, while some even stated that they expected something like this since they were in a study setting. We concluded each session by asking for the participants' secrecy on the content of the study because it would negatively affect the results of the study if other participants knew how to use the app, which would render the usability evaluation useless. Making transparent why it is important not to disclose the study content to other participants was successful, since we received feedback from multiple participants that they asked other participants beforehand what to expect, but were turned down because of our request.

10.4.1.2 Participants

We recruited 35 participants for our study from the university environment, by advertising our study on the campus and offering a link on a QR-Code for self-registration for possible time slots. The ages of the participants ranged from 20 to 42 years, with a mean age of 24.6 years, a standard deviation of 4.07, and a median of 23 years. Nine of them identified as female, 26 as male. Although the subject areas of the participants covered a wide variety of topics, this led to a bias in the sample due to the mostly young and educated participants (cf. Electronic Supplementary Material A.3.2). However, this participant group is similar to the main user group of smart speakers [1]. Interesting for our study is that the participants had different experiences with smart devices (cf. Electronic Supplementary Material A.3.3.1). Thus, we can examine differences between both groups of users, who on the one hand have experience with smart speakers and on the other hand do not have experience with smart speakers.

10.4.1.3 Materials

For our study, we built a setup to allow the participants to get used to a smart speaker and be able to compare the behavior of them with compromised and non-compromised smart speakers. The setup consists of the two Amazon Echo Dots with stands described in Sect. 10.3.2, a lamp with a smart light bulb and a smart plug connected to a Raspberry Pi to show the function of turning devices on and off via a smart speaker. Thus, a smart home environment should be simulated. A picture of the evaluation setup is shown in Fig. 10.4. Both smart speakers were required for our study, because in our scenario, one of the smart speakers was compromised, i.e., the simulated attestation of it would fail. Thus, we could examine how a user reacts

[1] https://speakergy.com/smart-speakers-statistics/

when finding out that one smart speaker is compromised. Additionally, we were able to observe and ask the participants how much trust they have in the compromised smart speaker and how that changes when the smart speaker is reset. After a press on the reset button, the attestation would succeed because the simulated malware would be removed by resetting the device. We provided the participants with a laptop to fill out the questionnaires for our study, which we prepared in a self-hosted instance of LimeSurvey.

Fig. 10.4 Picture showing the evaluation setup

10.4.1.4 Procedure

The study was conducted between 01/18/2023 and 01/31/2023. Each session consisted of several parts which were completed by a single participant in about 30 to 50 minutes. In the beginning, we asked participants for demographic information and four questions about their experience with smart speakers (cf. Electronic Supplementary Material A.3.3.1). All participants were given a short introduction, explaining the scenario and telling them about their tasks. Afterward, the participants were asked to interact with both smart speakers. In order to know what they could do with the devices, a handout with possible actions was provided. The described actions were regular smart speaker commands such as setting a timer or telling a joke. The actions we included to simulate a smart home environment were turning on and off a lamp, changing its color, and activating a connected device (a Raspberry Pi in our setup) via a smart plug. After the phase of familiarizing with the smart speakers, users were asked to perform an attestation. We alternated the versions of the app for each participant, i.e., every participant had either the transparent version or the standard version. First, we assigned the app versions alternately to the time slots for the study, but when some participants did not show up, we alternated the app versions in the order of appearance. In the study scenario, the participant chooses one of the smart speakers to be attested. This first attestation succeeds and the participant is asked to attest the second smart speaker. When attesting this second device, the attestation should fail. On the 'fail' screen, the participant is informed about the next steps. The first proposed advice is to retry the attestation in case of a transmission error. When the attestation fails repeatedly, the participant should reset the device using the button on the back of the device. After resetting the device, the participant is asked to attest the device again. This time, the attestation succeeds showing the successful reset of the smart speaker. After testing the attestation app, the participants answered the UEQ-S [270] giving us information about the user experience when using the app. Additionally, the questionnaire contains questions about the understanding of the process. We continued with a semi-structured interview for which we prepared some guiding questions to get more information about the users' perceptions. The study material can be found in Electronic Supplementary Material A.3.3.

10.4.2 Results

Since we used a smartphone application to evaluate the attestation mechanism, we wanted to ensure that the usability of our application was high enough to not influence our results. For this reason, we used the UEQ-S questionnaire. The results (cf. Fig. 10.5) show that our application has a good to excellent usability and thus does not influence our evaluation negatively.

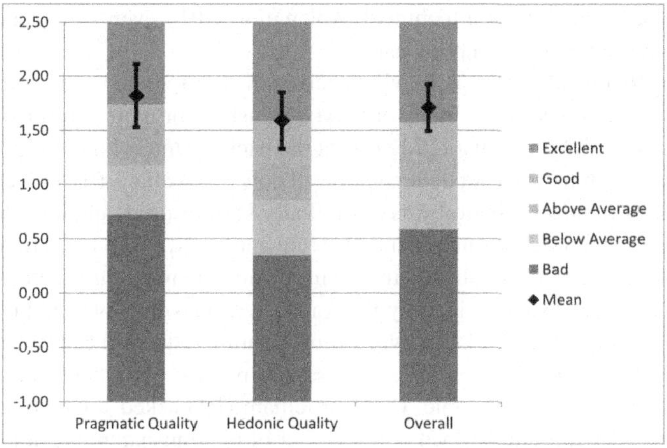

Fig. 10.5 Results of the UEQ-S questionnaire

10.4.2.1 Measuring Understanding based on the App Version

In order to examine the effects of the app version on the understanding of the attestation process, we analyzed the Think-Aloud study. There, we found evidence that many participants with the nontransparent application asked for more explanations on what to do. Participant P1 asked, for example: "Are there more explanations?" Similarly, P3 told us that we should implement an explanation in the app. P5, P13, and P28 asked for more information on "Why should I do the attestation process?", which shows that the meaning behind attestation was not understood. Some participants requested to be able to follow individual process steps better. For example, P9 wished "to see what the application tests at the moment" and said that "there were no explanations what I have done or how the attestation checks that the system is secure". Similar comments were made by P16, who asked for information "that explain what is happening and why" and stated that "I could see that something is happening but not what exactly happens" and P28 remarked "What is happening in the background and what do the tones mean?". Confusion was also expressed by P10, who stated: "Normally I am no fan of tutorials but in this case, a more obvious explanation would be great". P25 admitted "I only have a vague guess" about what is happening during the attestation process and wished for "a more transparent overview of the methodology of the attestation". P30 said "I have not understood what the application does" and wished for "small pieces of information on every screen", which describes our transparent version. P34 described the missing information as a reason for lack of trust in the process, as he is "not as convinced as he would be when having a technical explanation". P14 even concluded that the process was "not at all" comprehensible.

On the other hand, participants who received the transparent version of the application but did not read the explanatory text also asked for more information. Here, P4 stated: "I don't know what I did or what the background behind this process is." and P29 said "I think I did not understand well how this works". P12 reflected on the fact that the explanation button was not initially seen or used, and thus the attestation process could not be comprehended. More generally, P27 wished to have more information on why attestation should be used and explained that he knew what to do because of the texts provided by the transparent version, but could understand the process "only a little bit". The type of information requested can be categorized into more general information on the attestation process, and technical details for further information on the topic. One participant (P1) asked for both. Six of the participants with the standard version and four of the transparent version asked for general information on the process. Nine participants with the standard version and eight with the transparent version asked for further technical details.

We measured the understanding of the process in two different ways: The self-rating of the participants using questionnaire item B2 (*I understood how the process behind attestation works*) and the question in the interview to describe the process in their own words. Regarding questionnaire item B2, participants with the transparent version answered with a total score of 71 which is a mean score of 3.94 (4='Slightly Agree'), whereas the participants with the non-transparent version answered with a score of 59, resulting in a mean score of 3.47 (3='Slightly Disagree'). When the participants in the discussion were asked how well they could comprehend the process of the attestation, there was a slight difference between the participants with the transparent and the nontransparent version: We found discrepancies between the self-rating and the observed understanding of some participants: P4, P10, P14 and P29 claimed to understand the process in the questionnaire, but could not explain it in the interview. On the other hand, P19 underestimated her understanding in the questionnaire. While those who had little to no understanding at all were evenly distributed among the versions (6 transparent; 8 non-transparent), those who were able to freely reiterate how the process works, which was considered as high understanding by us, were more likely to be in the transparent version (8 transparent; 2 non-transparent). Participants with medium understanding were considered as those who had a general idea but could not explain the process in detail (4 transparent; 7 non-transparent).

10.4.2.2 Measuring Understanding based on the Explanation Text

For the examination of the influence of the explanation text on the understanding of the attestation process, we analyzed the interviews and compared the results to the self-reported understanding in the questionnaire. One remarkable finding is that some users who did not read the explanation wished for an explanation. P4 said that it would be great if he could read somewhere about how the process works. And P5 objected to the missing '?'-button for an explanation of functions. Similarly, P16 wished for an "i-button which explains what is happening and why". P33 liked that the app "is very simple", but would like "to have a possibility to get more information".

The positive influence of the explanations on the understanding of the attestation process was confirmed by several participants. For example, P11 read the explanation and reflected after conducting the attestation that before reading the explanation, he was not sure whether the attestation would work. This was confirmed by P21 who said "You could see that something is happening but not what exactly happens; with the explanations, it became clear what happens". P15 said that the process could be well comprehended and "I have a plan [of what is happening]". P17 elaborated that "there was an explanation button" and that the process was "explained well". In addition, the suitability even for technology non-experts was noted. In this regard, P17 elaborated that the process "seems to be plausible, also for people who have not a high affinity for technology" and it is "quite foolproof". P18 explained that the process could be comprehended well and that "the idea could be understood". P19 said that "the explanation was very good, short but detailed and not overloaded" and "I could follow it [the explanation] as a student who does not study computer science". P32 stated "I think the explanation is good. I am not so familiar with the devices. I study politics. I think it is well explained for people who know almost nothing [about IT-Security]." P22 who only read the explanation after conducting the attestation (using the nontransparent application) said "I would have understood the process better if I read the explanation first". P23 confirmed "it was nice that there was an explanation; without that, I would have pushed a button without knowing what was happening". Thus the process could be "100% comprehended". P35 said "I understood how it works; the explanation was fairly clear". Several other participants (P15, P17, P18, P35) concluded that because of the explanations, the process as well as the general idea behind the process could be well comprehended.

Similar to the analysis of the app version's impact, we analyzed those who asked for more information in the app. Out of ten participants who requested general information on the process, only one read the explanation, while nine did not. Among the 17 who wanted further technical details to look for more information on the topic eleven read the explanation and six did not. Regarding questionnaire item B2

(*I understood how the process behind attestation works*), participants who read the explanation answered with a score of 74 which is a mean score of 4.63 (5='Agree'), whereas the participants with the nontransparent version answered with a score of 56 resulting in a mean score of 2.95 (3='Slightly Disagree'). However, when the participants were asked in the discussion how well they could comprehend the process of the attestation, there was a clear difference between the participants who read the explanation and those who did not. In total, 16 participants read the explanation and 19 did not. 13 of those who had little to no understanding at all did not read the explanation, while only one of the participants who read the explanation could not comprehend how the attestation process worked. Furthermore, six of those who did not read the explanation and five of those who read it had a medium understanding of the process. Most significantly, ten of those who read the explanation were able to freely reiterate how the process worked, while none of those who did not read the explanation were able to do so.

10.4.2.3 The Connection of Trust and Understanding

Please note, that the evaluation of the perceived trust and the understanding is indirectly affected by the two different app versions and whether the explanation was read since these conditions affected the understanding. In the following, we will not distinguish between the versions since these effects are already covered above.

The participants who had specific expectations based on prior knowledge are the first to show that understanding affects trust. For example, P18 and P25 knew that training data is important to train language models but were skeptical of smart speakers because of privacy violation incidents that were made public. This indicates that the level of trust is influenced by prior knowledge and thus can also be affected by understanding obtained during the use of a security feature. During the interviews, we found that there are different types of trust which were influenced by the understanding of the process: On the one hand, we observed trust in the process and the belief that this technical measure can indeed detect malicious modifications of the smart device. On the other hand, we detected that understanding of the process resulted in disappointment because the participants lost their naive and illusionary trust in the device: Before learning how attestation works, they believed that this would also prevent the manufacturer from collecting data via the smart speaker. Learning that attestation 'only' prevents malicious third parties from (mis-)using the device reduced their illusionary trust which originated from their former perception of what protection means and who was perceived to be the antagonist. In total, 17 participants of our study stated that their handling of smart speakers would not change because of the attestation. The main reason for this answer is that the users' problem is not that the smart speaker could be hacked but that the personal and spoken information is sent to the manufacturer of the devices (P7, P9, P22, P34). Since attestation can only detect that a device is compromised but not whether the information is sent to the manufacturer or not, attestation does not solve this basic skepticism. These participants were slightly disappointed when they learned that the attestation process *only* guarantees that no malicious third party has access to the device and that attestation cannot limit the amount of data that is transmitted to the manufacturer. Therefore, transparency does increase trust in the attestation process but weakens the illusionary trust in misconceived security and privacy. This is also one of the reasons why participants indicated that they do not share some personal information with the smart speaker (P8, P15).

General trust issues with companies were brought up by several participants: P12 said: 'I lose the control over my data and do not know what the companies actually do with it'. The struggle to minimize the amount of data that is given away is described by P28: 'I try to minimize tracking in my life. I make the three extra clicks to decline tracking in cookie banners and get angry with my friends when they click on 'accept all'. I studied computer science and am aware of the risks associated with data collection. However, sometimes you get ads for something you were talking about a few days earlier and then you know: somewhere data was collected'. The fear that companies may monetize the collected data was shared by P5, who stated the concern that people can be manipulated to buy products. On the other hand, P1 had little concerns about giving away data: 'I know that companies get my data but I am interested in the technology. Furthermore, I am young and have nothing to hide'. Another factor for indifference towards data collection despite lacking trust in companies was stated by multiple participants: Resignation. After experiencing that their data already was disclosed to companies, many only could 'trust' the companies that their data was not misused. 'Tracking happens everywhere: via smartphone, browsers, or social networks. This could provide benefits to save time but then you get a suspicious offer for things you may need and then it gets creepy. It all depends on what is actually done with our data by the companies' (P30). 'I am aware that companies gather my data. It is the same issue with my smartphone' (P33). This attitude was shared by P11. P29 continued: 'Google has my data anyway via my phone. You can't do anything against it'. Similarly, P27 stated: 'They know all of us already. As soon as you start using the service of one company you lose all inhibitors. In my social environment, I frequently hear: 'Google already has my data thus I do not care anymore about my data''. A similar statement was made by P9. However, even those who were not concerned about the data collected by companies showed signs of increased trust in smart speakers after learning about attestation: For example, P17 does not think that Amazon cares about what he talks about in his living room, but still finds the concept of a device listening to his voice a bit creepy: 'You do not know who else is listening'. Therefore, attestation is useful to increase the trust of users who think similarly.

When investigating the trust of the participants towards the security provided by attestation, it became clear that the mental models of the users played a major role. These depended on how the participants imagined the motives and possibilities of different actors. P3 and P7 thought that hackers were no threat to them because they were not interesting enough to be targeted. P21 shared this belief, but still wanted to take action to prepare: 'I do not believe that I get personally hacked, but I do want to secure my data'. P35 distinguishes between different devices to form a threat model: 'Regarding privacy invasion I am more afraid of my smartphone than a smart speaker. The smart speaker is only in one room, but my phone is always by my side'. And finally, many still were convinced that data collection by companies is a bigger threat than malware and hacker attacks: 'I am not sure whether a hacker attack is the biggest problem. I am more afraid that a company misuses my data. [...] Big companies gathering my data is more scary than hackers' (P35). P24 underlined the perceived threat by data collection: 'I think a bigger threat than a private hacker is the power amassed by big companies. This could pose a threat to society'. The evaluation of different actors and their capabilities was summarized by P21: 'I am not concerned that secret services are getting data from smart speakers. I got a smartphone after all. I am not afraid of governmental surveillance, but I do not want companies to use my data for marketing; My biggest fear is not a specific 'hacker organization' but security breaches in general'.

These differences in the perceived power of different actors are similar to the misplaced illusionary trust in attestation. Users have specific mental models and these influence their threat models and the level of trust they place in someone or something. Therefore, prior knowledge is a major factor in evaluating threats and trust. This can be achieved by transparency during the interaction with a tool or by prior experience as P34 illustrates: 'I turn a blind eye to the influence of big companies in my everyday life. I once requested my data from Google and they had data from my childhood. Until now nothing bad happened to me, but it depends on how much you trust the company'.

10.4.2.4 Trust in Formerly Compromised Devices

Regarding our third hypothesis, we examined (1) whether device handling would change due to the possibility of attestation and (2) whether the interaction with the device hacked in our study would change after successful attestation following a reset (see guiding questions (iv) and (vi) in Electronic Supplementary Material A.3.1). Regarding the first question, six participants stated that the attestation increased their trust in the smart speaker (P5, P20, P21, P23, P25, P28). Although they described an improved trust in the device, some users stated that the handling of the smart speaker still would be the same (P20, P21, P25). Three users reported that the attestation leads to an increased sense of security (P6, P16, P35). The greatest influence of attestation was on five participants, who stated that the possibility of attestation would be a reason to purchase the device, although they do not currently own a smart speaker because they do not trust them (P11, P12, P18, P19, P28).

For the second question, 17 participants responded that a formerly compromised device would not be handled any differently than a non-compromised device after a reset and subsequent successful attestation. For participant P4, the reset with a successful attestation would increase the trust in the device. Seven participants stated that they would be more skeptical about the formerly compromised device and would do the attestation more frequently. Four participants would ask themselves how the smart speaker could be compromised (P5, P6, P22, P33). For six participants the compromised device would lose their trust and would not be used anymore. Two of those would also lose trust in similar devices, even if they had not been compromised (P1, P7). In conclusion, attestation influences the trust and sense of security positively and can also be a reason to buy a device. One participant even stated that he would pay extra for a device with a built-in attestation function (P4): In case of a device costing €50, he would be willing to pay €10–15 extra for the attestation function. Otherwise, a fundamental skepticism towards smart speakers and their manufacturers cannot be changed even by attestation. However, most of the participants trust the attestation when a formerly compromised device is attested successfully after a reset.

10.5 Discussion and Conclusion

In this section, we reflect on the results to test our hypotheses and answer our research questions.

10.5.1 Findings and Implications

First, we examined the difference in understanding of the attestation process between users of the transparent version of the app and those using the nontransparent version. During the interviews, we observed that participants with the nontransparent version of the app asked more frequently for more information on the process of the attestation. However, this might be an effect of the sample size, the design of the versions itself, or the expectations of the participants. The latter can be derived from the quotes of P5 and P16 who asked for a '?'- or 'i'-button for more information. This implies that users do not expect transparency in each step but a central source of information. In the interviews, we could observe that participants who had the transparent version of the app had a slightly better understanding of the process than those who received the nontransparent version. Therefore, we formulate the hypothesis that the transparency difference in the app version does not have a relevant influence on the understanding of the attestation process.

Second, we analyzed the difference in understanding of the attestation process between users who read the explanation text and those who did not. We could show that the majority of those who wanted general information on the process did not read the explanation, while the majority of those who read it requested further technical details to dive deeper into the topic. Additionally, we could confirm that all participants who were able to reiterate how the process works in the interview had read the explanation. Overall, we formulate the hypothesis that users understand the process of attestation better when reading the explanation text in the app.

Therefore, we show that making security mechanisms transparent for end users supports their understanding of the overall mechanism. This combats the problem of users not understanding how their data is processed, as stated by Alshehri, Spielman, Prasad, and Yue [261] and the practice of designing smart speakers as black-box devices [210]. Therefore, we can answer our first research question (**'How does transparency affect the users' understanding of remote attestation?'**) with: Increased transparency led to an increase in understanding in our sample.

When investigating how understanding and trust are related, we found that the mental models of the participants played a major role. Those who had experience with security issues or heard of privacy violations were more skeptical than others. Furthermore, many participants thought that they were not interesting enough to be targeted by hackers. However, in the context of IoT, this is not the correct threat model. Users are more likely to fall victim to broadly deployed ransomware or botnets than being targeted as individuals by hacker groups. Therefore, educating these users is highly important to achieve actual security. By utilizing transparency the understanding of users can be enhanced, which proved to be essential for building trust in the security mechanism and to reduce false mental models such as misplaced trust. This illusionary trust was placed in the attribution process due to false mental models and by attributing properties to the system due to lacking understanding.

Therefore, we can answer the second research question (**'How does the users' trust change by gaining a deeper understanding of remote attestation?'**) with: Understanding affects the mental models of the users and enables them to estimate the correct level of protection. Furthermore, false attributions are revoked and real trust can be built.

Our third sub-goal was the investigation of whether participants would trust a formerly compromised device after it was reset and successfully attested. During the interviews we received mixed feedback: On the one hand, resetting and attesting a corrupted device increased the trust of some participants. On the other hand, some participants indicated that they would become suspicious of all similar devices if they discovered a corrupted device in their vicinity. Most participants agreed that the possibility of attestation does not affect how they interact with IoT devices in general. Some indicated that their trust is strengthened, but they would not share more information with the smart speaker than they would without attestation. Even if the device was compromised and restored, the behavior towards the device would not change. However, participants stated that they would be alarmed and conduct the attestation process more often after the discovery of a corrupted device. Some stated that they would test the device once a day for at least some weeks after regaining trust in the device. Others stated that they would stop using the smart speaker if it was compromised again after the reset. This aligns with findings from literature pointing out that security is an important factor for users to trust IoT devices [249]. With these findings, we can answer our third research question (**'Do users trust formerly compromised devices, that have been restored and verified with remote attestation?'**) with: Yes, most users do trust formerly compromised devices if they have been restored and verified via remote attestation.

Overall we could investigate that transparency indeed increased the trust in remote attestation. By giving the users the possibility to comprehend the process, they were able to evaluate the result of the attestation themselves and could take action as they deemed necessary. The fact that most participants of our study would use a formerly compromised device after reset and successful attestation shows the level of trust towards the attestation.

10.5.2 Limitations and Future Work

Our study is subject to several limitations. First, our sample consists of relatively young and well-educated participants. Although this resembles the target group of smart speakers closely, it still makes our study not representative of a broader target group. Second, our methodology is a qualitative study. While qualitative studies allow for a more in-depth investigation of a topic, they are not representative on a large scale. By reaching saturation during the interviews due to the lack of new input from the participants, we can claim that most perspectives are represented. However, even though we interviewed participants of different nationalities, a sample from a different country may bring up other results since risk cultures differ based on the background of people. For example, Reuter, Kaufhold, Schmid, Spielhofer, and Hahne [271] showed that Germany had the strongest privacy concerns of all investigated European countries when interacting with social media. Therefore, we encourage other researchers to use our study design as a basis to conduct comparative studies or even quantitative studies on this topic in future work to broaden the perspective of the research community on this topic. Another limitation is the suspicion of some participants that the smart speakers were not really compromised. This indicates a different perception of the situation in this study compared to the real world which could result in a different behavior in the real world. Future work could also develop other methods of transparency to evaluate their effect on the trust of users towards attestation. Another interesting approach would be testing the limits of transparency. In our study, we found that transparency enhances the understanding of the users and thus the trust in the attestation process. The results of the usability scale and the fact that the participants could reiterate more or less how the process worked led to the conclusion that our setup provided a useful amount of information. However, including too much information could lead to information overload and thus reduce the level of understanding. Testing the limits of the amount of information which is helpful and the point when it becomes detrimental could greatly enhance the research on information overload and decision fatigue.

10.5.3 Conclusion

Finally, we can summarize our findings to answer our main research question: **Does transparency enhance trust in remote attestation?** In our study, we confirmed that transparency as an HCI intervention does increase the understanding of processes such as remote attestation. We were also able to confirm that understanding the process of remote attestation leads to increased trust in the overall process. One example of this is the audio communication between the devices. Most participants thought it would just be feedback for them to show that the attestation is in process. However, by realizing that audio communication is useful to avoid remote attackers, users did trust the attestation process more. This trust even allows them to rebuild trust in formerly compromised devices which were restored and verified by remote attestation. Therefore, we conclude that trust is based on prior knowledge as well as mental models. Transparency affects both by increasing the understanding and thus the knowledge about a security mechanism, and on the other hand it reduces illusionary trust which originates from mental models and false attributions. Thus, transparency helps build real trust in the actual functionality of security and privacy features. Therefore, transparency indeed supports users to calibrate their trust to the trustworthiness of the system. Therefore, we can answer our main research question in the affirmative, and conclude that transparency enhances trust in remote attestation.

Acknowledgments This work was supported by the Deutsche For-schungsgemeinschaft (DFG, German Research Foundation) under Grant 236615297 (SFB 1119 CROSSING); and this research work has been funded by the German Federal Ministry of Education and Research and the Hessian Ministry of Higher Education, Research, Science and the Arts within their joint support of the National Research Center for Applied Cybersecurity ATHENE.

Paper F: Decision-based Data Distribution (D^3): Enabling Users to Minimize Data Propagation in Privacy sensitive Scenarios

11

Abstract

In many scenarios, users have to communicate sensitive data with third parties such as doctors, lawyers, insurance companies, social workers, or online shops. Handing over personal data is necessary to use those services, but delegating tasks to increase efficiency still poses the risk that personal data might be leaked. To minimize this risk and further enhance the privacy of users, we propose an interaction concept that uses layered encryption of messages to provide a trade-off between privacy and usability. Users can choose which data is additionally encrypted in an inner layer, e.g. only for the eyes of their doctor, and which data is available in an outer (encrypted or unencrypted) layer for all staff members. Another benefit is the hiding of sensitive data from package inspection or crawling algorithms via emails, while less critical parts can still be processed by these systems via the partial access. To investigate this concept, we derive relevant use cases for form-based communication via email from a quantitative pre-study

Original Publication *S. Linsner, K. Demuth, M. Fischlin, and C. Reuter, "Decision-based Data Distribution (D^3): Enabling Users to Minimize Data Propagation in Privacy-sensitive Scenarios,"* Proceedings on Privacy Enhancing Technologies (PoPETs), *no. 4, 2024*, ISSN: *2299–0984*. DOI: https://doi.org/10.56553/popets-2024-0113. *[Online]. Available:* https://petsymposium.org/popets/2024/popets-2024-0113.php

Supplementary Information The online version contains supplementary material available at https://doi.org/10.1007/978-3-658-47445-4_11.

S. Linsner, *Privacy Preserving Data Management*, Technology, Peace and Security I Technologie, Frieden und Sicherheit,
https://doi.org/10.1007/978-3-658-47445-4_11

with 1011 participants, showing that general practitioners are the most suitable use case. We developed demonstrators for this use case and evaluated them in a qualitative study with 42 participants. Our results show that the possibility of minimizing the propagation of sensitive data through additional encryption is highly appreciated and the usage of form-based communication is a promising approach for digital transformation.

11.1 Introduction

In December 2022, Google announced it would integrate client-side encryption in its mail clients for enterprise workspace and educational customers. While this clearly advances end-to-end security for users, it also comes with a downside, as Google notes on its support page[1]: "Client-side encrypted files and emails aren't scanned for phishing and malware, because Google's servers don't have access to the content." Hence, while users gain confidentiality through the established end-to-end encryption, they also degrade their own security by precluding Google from scanning for malicious emails.

The trade-off between confidentiality and accessibility is not new, and has been discussed extensively in many facets before [272–274]. The reasons for giving access to protected communication, where today this protection is either accomplished via plain encryption or by secure channels like TLS, may be conflicting security desires [275]. The end-users may wish for malware protection or intrusion detection, as in the Google example above, or may have other, more subtle reasons like caching or compression for performance [276].

Arguably, the most visible example where confidentiality and accessibility clash is lawful interception. Recently, the most prominent case may have been the efforts of major providers to integrate detection mechanisms for illicit content, such as child sexual abuse material (CSAM) [277]. This includes Apple's NeuralHash, Microsoft's PhotoDNA, and Facebook's PDQ, which scan for suspicious image or video material at the client side and report potential cases to the providers. Beyond the political dimension, it turns out that the approaches behind the deployed perceptual hash functions, which should tolerate small changes in the images, are all cryptographically weak [278, 279] and thus do not provide the required level of effectiveness.

From a technical viewpoint, the situation in the area of intrusion detection and malware protection for encrypted data looks much better than for perceptual hashing.

[1] https://support.google.com/a/answer/10741897?hl=en, accessed August 21st, 2023.

Initially, the most common solution for intrusion detection was to give the firewall access to the shared key so that the system could access the protected data in clear form. Alternatively, but effectively identical, the middlebox may securely connect to each end-user and relay the data. In recent years, researchers have developed improved solutions for this problem, aiming to re-establish privacy for the users and their data while maintaining a minimal level of accessibility for the detection systems. Examples are the Blindbox protocol [280], the recently proposed concept of zero-knowledge middleboxes (ZKMB) [281, 282], and multi-context TLS [283]. Roughly, these approaches encrypt communication with a secure end-to-end mechanism but also provide additional information to the detection system, enabling it to check for malicious content.

Another recent proposal is to use stealth key exchange and channels in TLS 1.3 [284], where the sender determines to which partial content the detection system has access, and which part should be kept confidential. The solution relies on the ability to generate a second shared key in the key establishment of TLS "for free". Consequently, one key can be shared with the middlebox, and the other is only known to the end-users. The sender can then decide which data should be encrypted additionally under the extra key and thus remain hidden from the middlebox. The middlebox only gets access to the other data. We note that outsiders do not have access to either kind of data. Therefore, in contrast to Blindbox and ZKMB, the approach reveals partial information to the detection system but is faster and easier to integrate into existing network protocols. In the context of this paper, we use the term partially-accessible encryption to describe the utilization of layered encryption to hide sensitive information from recipients without the second shared key.

Given that we now have the technical possibilities to select a continuous balance between confidentiality and accessibility in encrypted communication, the question remains what the users want and if they would be willing to adopt such a system. Therefore, our research question to investigate possibilities, trade-offs, and pitfalls for layered encryption in real-world scenarios is:

> **'How do users perceive the possibilities of partially-accessible encryption in everyday scenarios?'**

In order to answer this question, we first conducted a representative quantitative study in Germany with 1011 participants to find out which communication is especially perceived as sensitive by users and, therefore, appropriate for an idea like ours. We found the communication with doctors and lawyers to be particularly sensitive and chose the former for the implementation of demonstrators. We opted for two demonstrators to investigate the so-called control paradox, a phenomenon described in the literature that leads to increased data disclosure when granting more control over the disclosure to the users [34] [35]. Therefore, we built two different versions of a contact form which can be used to request a prescription or a referral from a doctor. In one version, the user can decide which information is visible only to the doctor. In the other version, all information is visible to the entire medical office, with the exception of one text field in which the user can write information that can only be seen by the doctor. These two demonstrators were finally evaluated in a qualitative study with 42 participants. We found that around 74% of users preferred the version where they can decide which information is visible only to the doctor and appreciated the freedom of choice granting them sovereignty over their data. The participants highly appreciated the easy-to-use concept of partially-accessible encryption and were curious about a real-world adoption since our concept would reduce physical and social barriers to seeking help and would make interaction with doctors easier. Furthermore, the majority of the participants appreciated the feeling of control over their own data.

11.2 Related Work

As this paper covers multiple research areas, related works to the presented concept feature discussions about secure communication of sensitive data in general, (partially-)accessible as well as usable encryption. Lastly, the discussion about the control paradox has to be considered for this study and is therefore outlined.

11.2.1 Sensitive and Secure Communication

The research on secure communication with professions where sensitive data needs to be transmitted has widely differing approaches.

Lerner, Zeng, and Roesner [285] specifically targeted the communication of sensitive data via email and introduced an email client to facilitate email encryption. Another very widespread approach to secure communication is the use of blockchain, e.g., to store and share electronic medical records securely [286]. Li, Lal, Conti, and Hu [287] used blockchain technology to propose a scheme for evidence management to achieve witness privacy. Zheng, You, and Hu [288] used blockchain for secure communication of medical insurance claims to ensure privacy and legitimacy. Yadav, Charles, Pandey, and Gupta [289] similarly addressed the problem of privacy-preserving insurance registration with blockchain and smart contracts. A different approach to secure communication, only prevalent in the e-commerce domain, was the use of quantum security. Most recently, the work of Thapliyal and Pathak [290] presented four different quantum protocols to guarantee security in online shopping, building upon approaches by Chou, Lin, and Zeng [291] and Huang, Yang, and Jia [292].

11.2.2 Partially-Accessible Encryption

Governmental organizations argue the importance of so-called lawful interception, usually with the need to fight crime or for reasons of national security ("going dark"). However, beyond the political dimension of this, the examples of client-side scanning for illicit material, as in the case of NeuralHash, PhotoDNA, and PDQ, show that the technical solutions in this area are far from mature [278, 279]. The European Union is currently discussing similar measures under the term "chat control". The EU proposal has received a lot of criticism regarding data protection, and its lawfulness is highly disputed, even within the EU [293].

The possibility of middleboxes being able to access encrypted data legitimately has a long history. It is based on the observation that it may be in the interest of users (e.g., when protecting them from malicious attachments or when compressing data to save resources) or in the interest of the party running the detection system (e.g., a company trying to prevent attacks on their system) [294]. We focus here on cases where the user can decide whether to use these measures or not. In the widely deployed TLS protocol, granting the middlebox access is possible via the usage of static Diffie-Hellman keys (or, in earlier versions of TLS, also for RSA keys). In this case, the middlebox would be able to decrypt the secured communication data. Unfortunately, this approach with static keys infringes on the notion of forward security, that past sessions should still be protected if the adversary gets access to static private keys. Yet, Green et al. [295], for instance, describe such a TLS 1.3 variant with static keys. Another widely used option is to let the middlebox break up end-to-end encryption. That is, the middlebox establishes a secure connection with one user and another one with the other user, re-encrypting data when routed through the middlebox. This is, for example, the path middlebox-aware TLS (maTLS) takes [296]. In addition to the loss of end-to-end security, these proxy solutions introduce other risks, such as insufficient certificate validation [297].

As mentioned in the introduction, the shortcomings of the above approaches lead researchers to develop advanced methods such as Blindbox [280], ZKMB [281, 282], multi-context TLS [283], and stealth key exchange for TLS 1.3 [284]. However, Fischlin [284] was able to integrate the key exchange into TLS 1.3 without modifying the network layer of the original protocol, although lacking in flexibility and security compared to Blindbox and ZKMB. The idea by Fischlin [284] of sanitizable channels to allow for a middlebox to alter parts of the payload builds on the concept of sanitizable signatures [298], sanitizable signcryption [299] and access control encryption [300–303]. Fischlin [284], however, applies this concept to symmetric-key cryptography in real-world channel protocols, which had not been covered before.

There are some methods that follow a similar objective as this paper, namely, structured encryption and, more specifically, searchable encryption and variants of both. They describe ways to store data and perform actions, e.g. by searching through it via an untrusted server, while the server cannot learn any information about the data [304–311]. Most research, however, sees the use case for this in the storage of large databases [312], wherein the storing and accessing of data is done by the same person [313–317], contrary to the approach presented here.

Finally, layered encryption is also used in onion routing protocols, especially Tor [318]. There, the goal is to hide path information from the onion routers and outsiders. This is accomplished with an overlay network, whereas the other aforementioned solutions for detection systems usually try to stay close to the original network infrastructure. Ideally, the content in onion routing protocols is still end-to-end encrypted and only available to the intended recipient.

11.2.3 Usability of Encryption

There has been a general discussion about the security usability trade-off for some time. Some authors agree that there is a clear trade-off between security and usability [42–44] and try to mitigate the effect in differing ways, trying to aid practitioners [319, 320]. Others see the current discussion as less expedient and point out that a more fine-grained assessment of a possible security usability trade-off is necessary. They state that uncertainty about factors leading to the problem and the methods used to research the suggested dichotomy are insufficient for a clear statement [321, 322]. Some, on the other hand, claim to have found ways to diminish the caused harm [323, 324]. To consider a possible trade-off, this work follows the proposal of Alsuhibany [44] to analyze the usability of a new security feature precisely. Other studies have focused more on concrete practice examples regarding the usability of encryption tools and mechanisms. Since Whitten and Tygar [325] has shown that the usability of PGP 5.0 was insufficient, there has been a discussion about usable email encryption. Multiple papers showed similar results to Whitten and Tygar [325] by conducting user studies on usable email encryption by means of PGP clients, stating that users would be frustrated or unable to successfully use these or giving recommendations for future improvement [326–330]. Borradaile, Kretschmer, Gretes, and LeClerc [331], on the other hand, were able to show that if the motivation was high enough, over 50% of study participants would continue to use PGP for secure email communication after a short training period. Since the end of 2022, Google has been offering the possibility of email encryption to selected customers (Enterprise Plus, Education Standard, and Education Plus) in the browser-based mail client. The confidential mode introduced previously in 2018, although perceived by users to use end-to-end encryption, did not offer this service [332]. The new client-side encryption (CSE) from 2022, essentially represents an end-to-end encryption and allows users to control the key management. Because this hinders Google servers from accessing the email content after encryption, emails can no longer be checked for malicious content.

11.2.4 Control Paradox

Literature often shows that it is important to give users control over their data [41]. Other than by technology, the use of data can also be controlled by laws or policies which need to adapt to the rapidly changing technology [333]. In contrast, recent literature shows the existence of the so-called 'control paradox': Brandimarte, Acquisti, and Loewenstein [34] and Boer, Herder, and Lieshoud [35] showed that users reveal more personal information when given more control about their shared personal information, which in turn leads to a higher risk of identifiability. However, Colnago, Cranor, and Acquisti [334] warn that self-assessments by study participants about their privacy concerns and behavior can be misleading and that respective privacy measures are often taken more regularly than participants think. Additionally, concerns about security and privacy vary widely among individuals and particular risk scenarios [334–336]. Therefore, we can also examine the control paradox and observe which information is published when users receive control over the disclosure of their personal information.

11.2.5 Research Gap

As discussed, allowing users to encrypt their communication can increase their security and privacy, but this requires some effort. This is the reason for lacking adoption of encryption tools such as PGP. Another usability-related problem is that users can make poor privacy decisions because of the negative effects of information overflow or the control paradox. Even if those problems are overcome, the user has to deal with the trade-off between privacy and usability. As the Google CSE debate shows: better encryption can diminish the possibility of malware checks. Furthermore, encrypted content is not available for search operations in the mailbox.

To combat these problems, we propose the usage of partially-accessible encrypted messages. The users are enabled to define which information is protected with additional encryption and which information is transmitted 'normally' (either unprotected or still encrypted but for a broader audience). This way, we reduce the cost of the privacy-usability trade-off and allow search operations on the accessible part of the message. By freedom of choice, users can opt not to use the feature if they do not want to invest time and thought for additional protection. On the other hand, users with a strong preference for privacy can fully customize the message to meet their desired level of protection. Enabling users to act without enforcing it makes this system viable for a variety of target groups. For this purpose, we provide a concept for the use of partially-accessible encrypted messages in an everyday use case (see Sect. 11.4) derived from a representative pre-study (see Sect. 11.3). To investigate the usability of such a system as well as the effects of the control paradox, we implement demonstrators for a qualitative study (see Sect. 11.5). An overview of the applied methods can be found in Electronic Supplementary Material A.4.4.

11.3 Quantitative Pre-study

In order to find suitable use cases for our study, we conducted a quantitative pre-study. This was considered necessary to ensure that the design and the use case were suitable for German participants. Literature shows that Germans differ from other European (and non-European) citizens with regard to risk culture and attitude towards authorities [271]. For a detailed discussion of these regional differences, see Sect. 11.6.4. Furthermore, the recent pandemic accelerated the adoption of digital tools for work and social life. Therefore, reliable data for an appropriate use case for our interaction concept is needed.

11.3.1 Methodology

While the personal data collected was limited to age, gender, education, income, and state, participants were transparently informed about the goals of the study and then asked for their informed consent to participate. For the inquiry, we selected GapFish (Berlin) as an ISO-certified panel provider, which ensures panel and data quality, representativeness for the German population, security, and survey quality within their panel of 500,000 active participants. After transmitting the final questionnaire, GapFish programmed and hosted the online survey. Once final quality checks, a soft launch, and mutual agreement were achieved, they invited participants from their panel to take part in the survey in April 2023 (N=1,011, t: 1 min. per question). The sample comprised 1011 persons living in Germany, of which 517 identified as female, 493 as male, and one as diverse. The age of the participants was distributed in a broad range: 8% were between 18 and 24, 15% between 25 and 34, 15% between 35 and 44, 16% between 45 and 54, 19% between 55 and 64 and 28% were 65 or older. The participants' monthly income was distributed as follows: 18% had an income lower than €1500, the income of 27% was between €1500 and €2600, that of 35% was between €2600 and €4500, and 19% had an income above €4500. 26% of our participants had at least a general high school leaving certificate (German average 28.6%), 5.49% had a former GDR general education certificate (German average 6,5%), 30.56% had an intermediate school-leaving certificate (German average 23.5%), 35.5% had college/university entrance qualification (German average 33.5%), and 16.29% had studied at a university or college (German average 17.3%). The values for the German average were taken from the federal office for statistics (Statistisches Bundesamt)[2]. For the sake of research economics we chose to participate in a joint survey with other researchers of our department (18 items total). We are aware that this might affect the answers of the participants, but since we are using this pre-study only for choosing a use case, the validity of our main evaluation (see Sect. 11.5) is not affected.

11.3.2 Items and Results

To find out which communication is particularly sensitive for users, we posed the following questions in the survey:

11.3.2.1 Use Case

To find the most suitable use case for our concept, we designed the first item: "How important is it to you to have a secure line of communication with the following people and institutions?" The users could answer this question by rating the importance on a 5-Point-Likert-Scale ranging from very important to very unimportant. The following communication partners were surveyed: clergy, chaplains, social workers, colleagues, supervisors, customer service from online shops, insurance companies,

[2] https://www.destatis.de/EN/Themes/Society-Environment/Education-Research-Culture/Educational-Level/Tables/educational-attainment-population-germany.html

friends, lawyers/courts, general practitioners, medical specialists, and family members. Not all of those are relevant for our use case, but it would be a bias to nudge the participants toward a desired result. Therefore, we include typical groups and persons that the participants are familiar with. For our use case, we will pick the highest-rated of the use cases that work for communication via an online form. The results for the first item can be seen in Fig. 11.1.

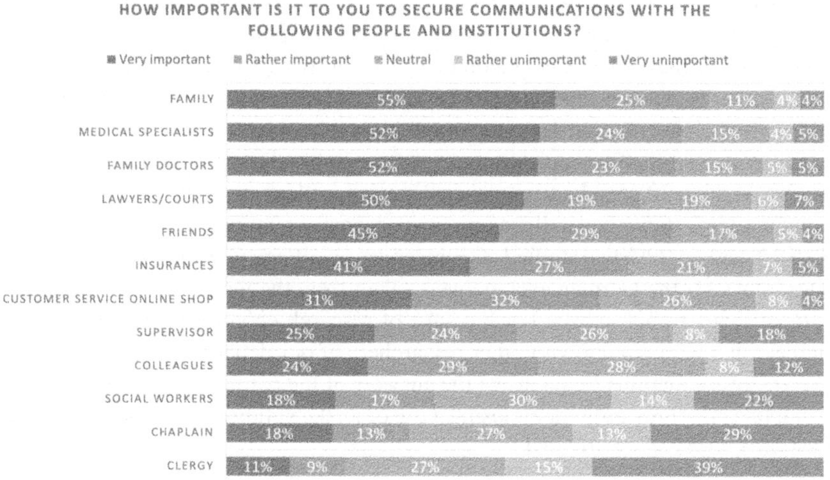

Fig. 11.1 Results of the first item of our pre-study

The communication most worthy of protection is the communication with family members, followed by communication with medical specialists, family doctors, and lawyers. Also worthy of protection, but not as sensitive, is the communications with friends, insurance companies, and customer service from online shops. The communication least worthy of protection is the one with supervisors, colleagues, social workers, chaplains, or the clergy. For our experimental form-based communication approach, it would have been difficult to build a demonstrator for communication with family and friends. Furthermore, it is unlikely that family members would communicate via an online form for matters of their daily life. Therefore, the remaining appropriate use cases were: medical specialists, general practitioners, and lawyers. Since we were planning to advertise our main study at the university, we chose general practitioners because it was to be expected that the participants attending our study would be most familiar with this use case.

11.3.2.2 Sensitive Data

To further investigate which type of data is most sensitive to users in communication with these actors, we designed the second item: "Which data would you disclose to third parties without concern?" The answers were "disclose" and "do not disclose" for the following data: First name, last name, address, email address, billing address, Customer ID (at customer support), telephone number, mobile number, IBAN, PayPal account, social media accounts, social media accounts of friends (advertise friends to get a bonus). The results are displayed in Fig. 11.2.

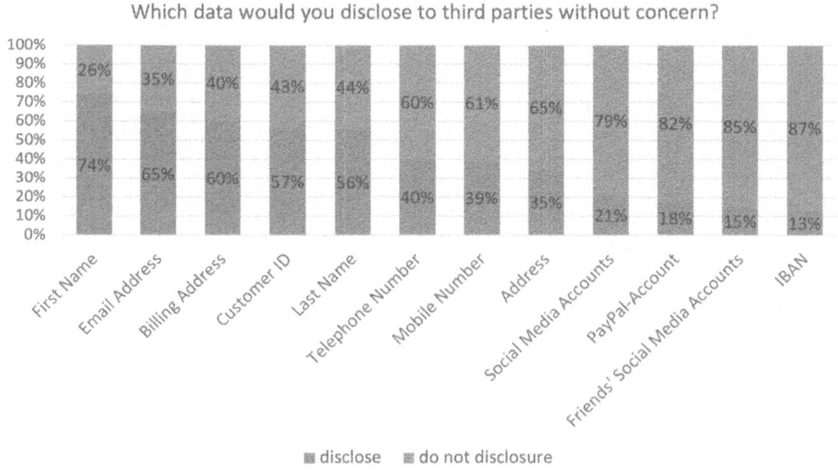

Fig. 11.2 Results of the second item of our pre-study

11.4 The D³ Interaction Concept

As related work has shown, there are several approaches to enhance the confidentiality and security of C2B communication. In this paper, we focus on the approach of Fischlin [284] since it offers additional features such as deniability of usage. The sTeaLS protocol [284], as explained in the introduction, allows the generation of two keys in a single connection. One key is shared between the endpoints and the detection system, representing a regular TLS channel with full access for the detection system to the data. The other key, called the stealth key, is exclusively shared between the endpoints. The former key protects the so-called TLS channel, whereas the latter key secures the so-called stealth channel within the TLS channel via pre-encryption of the sensitive data. This two-key approach enables the user to decide which parts of the encrypted data should be shared exclusively with the receiver and which parts may be accessible by the detection system and potentially administrative personnel. This allows for an interesting new perspective: Instead of communication with a faceless organization, the communication is now directly related to another person inside the organization. The person receiving the message is not someone who randomly got assigned to the task but is addressed directly. We call our interaction concept "Decision-based Data Distribution (D^3)". Note that outsiders do not have access to any transmitted data. The concept is illustrated in Fig. 11.3.

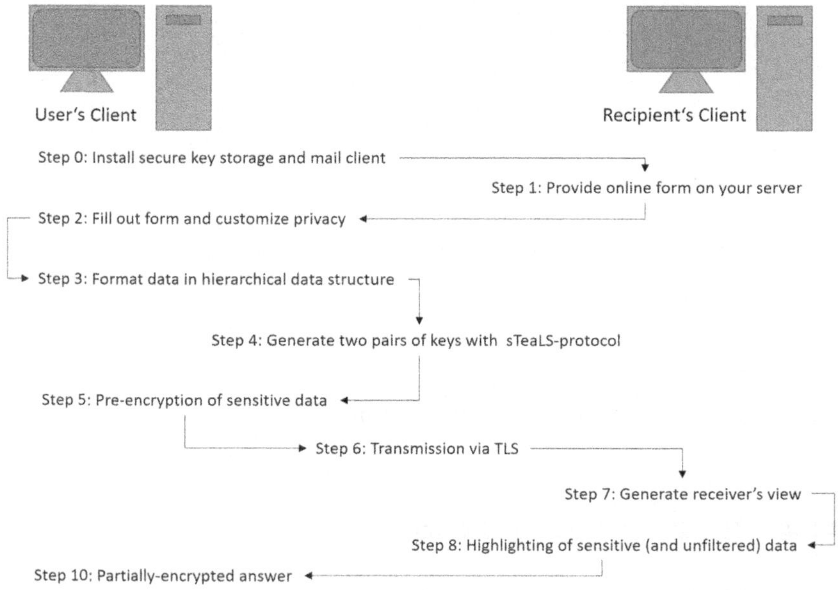

Fig. 11.3 Illustration of the D^3 concept: Interaction between user and recipient

The described approach is independent of the use case. However, our study will focus on one particular use case derived from the pre-study (see Sect. 11.3). Our design comprises an online form, a key wallet, and TLS connections for data transfer. After the initial contact via the online form, the correspondence is via email, in which the form structure is still represented. We assume that the 'final' implementation consists of an application similar to a mail client or an add-on for an existing mail client which contains the users' key wallet and has an authentication mechanism to identify the user in the decryption process. Furthermore, we assume that everything is implemented correctly and keys are stored securely. The receiving party provides an online form (e.g., on their website) with input fields for necessary information needed to fulfill a service. The users fill in the required information and define which information should be considered sensitive and, therefore, which information should be sent over the stealth channel. After submission, the structure and content of these forms are converted into a structured data format such as HTML, XML, JSON, or similar. Sensitive data, which is marked for stealth communication, is marked as 'secret' in the data structure. Furthermore, the sTeaLS protocol is initiated, and the TLS and stealth keys are computed. The sTeaLS key is added to the user's and recipient's key wallet. All 'secret' data is encrypted with the stealth key. After encryption, 'CONFIDENTIAL' is added to the result as a flag for regular TLS decryption. Additionally, if a 'secret' field is empty, it is filled with the string 'no input', which is then encrypted with the stealth key to provide indistinguishability if secret data was submitted or not. The transmission itself is similar to regular TLS. On the receiving side, the form is displayed, and if the viewer possesses the stealth key, the secret information is decrypted and displayed because the flag 'CONFIDENTIAL' is revealed in decryption. Otherwise, the message 'access denied' is shown in the secret sections of the form. An optional feature is the highlighting of the confidential content for the recipient. This serves two purposes: (1) the recipient is aware of which information should be treated with care to protect the sender, and (2) the recipient knows which content was not checked by the firewall and, therefore, should be treated carefully, since it may contain malicious links or other forms of social engineering. To utilize the full potential of the stealth channels, the recipient's response is also partially encrypted. Therefore, information that was considered sensitive by the initial sender and marked as confidential is not sent in plain text in the recipient's response. The partially encrypted mail ensures that the sensitive information is not leaked by unauthorized access to the sender's computer (e.g., by persons in the same household). Due to the form-based communication, the confidential fields can be hidden by default. The same mechanism that prevents other people on the receiver's side from viewing confidential data prevents users of the sender's computer from viewing it. To view the encrypted parts of the emails, the user has to be authorized by the client that hosts the key wallet. Then, the user can manually decrypt the confidential fields. The purpose of this is that sensitive information is never stored persistently on the computer and is only temporarily available to be displayed.

Please note that this concept relies on compliance by the involved parties, like most other security- and privacy-critical applications that involve human beings. Circumventing mechanisms or sharing credentials render those concepts useless. Therefore, we propose this concept for companies and individuals that want to provide increased privacy for their customers or clients. Especially for jobs that fall under the obligation to secrecy, like the medical confidentiality of doctors, the secrecy of confession for religious matters, or the professional discretion of lawyers.

11.5 Qualitative Evaluation

We built and evaluated demonstrators to gain user feedback on our interaction concept. The results are used to evaluate the practicability and to enhance our concept. A qualitative approach was chosen to allow us to watch the participants interact with our demonstrators and gain first-hand impressions. Furthermore, the qualitative evaluation provided us with in-depth feedback and discussions of our concept.

11.5.1 Methodology

11.5.1.1 Participants

The participants were hired from the university environment. We advertised the study in lectures and via a mailing list for people interested in our user studies. We recruited 42 participants in total, 10 of whom identified as female and 32 as male. The participants' ages ranged from 17 to 42, with a median of 22.5 and a standard derivation of 4.61. Before conducting our study, we obtained ethical approval for our study from the ethics committee at our university. This ensures that we adhere to the university's ethical guidelines regarding study design, data storage, and processing. Accordingly, every participant signed an informed consent form describing the procedure of the study, the data storage, and processing. Every participant was compensated with €20 for participating in the study and informed that they could leave the study at any point without having to provide a reason. We removed the pseudonyms of the participants from results that cover sensitive use cases to enhance the anonymity of the participants.

11.5.1.2 Materials

We decided to use mock-ups for our study because they are lightweight to implement, easy to adjust, and are not dependent on an internet connection. As explained in our concept (see Sect. 11.4), we assume that technical preconditions are given since those are out of the scope of this paper (e.g., wallets to store keys). For our main purpose, the user study that investigates whether people perceive the benefits of partial encryption when communicating with doctors, it is not necessary to implement a mail client, a secure key storage, and an administrative software for a doctor's computer. We can simulate all of these using interactive mock-ups without creating dependencies and possible bugs that would interfere with our study. Therefore, we created interactive demonstrators using Axure[3], which offer the participants of our study the possibility to experience our concept from the sender's and recipient's view. We chose to build two different demonstrators because we wanted the users to choose themselves which information should be protected. However, literature highlights the control paradox: users tend to reveal more personal data when they have more control over it. To investigate this, we propose two different versions for our demonstrator (see Fig. 11.4 and Fig. 11.5):

Fig. 11.4 Mock-up for the non-customizable version: online form (left), doctor's view (middle) and answer email (right)

[3] https://www.axure.com/

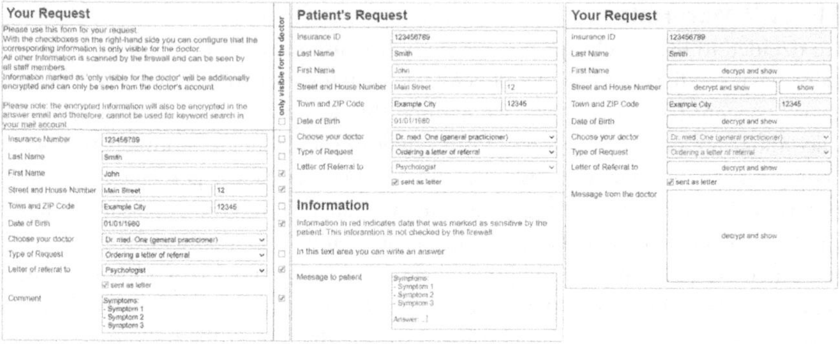

Fig. 11.5 Mock-up for the customizable version: online form (left), doctor's view (middle) and answer email (right)

(1) One version has no option for partial encryption besides one single text area that is, by default, encrypted with the sTeaLS key. This is the baseline version for the control paradox since no customization is allowed. For brevity, we refer to this version as the *non-customizable version* (see Fig. 11.4).

Each demonstrator consists of three views: (a) the online form, (b) the doctor's view, and (c) the answer email. We decided to include all three views in the demonstrators to give the participants of our study an overview of the effects of the privacy features.

The first view consists of three parts: A part for the patients' information to identify them, an information text, and the message to the doctor. The first part comprises several input fields to fill in general information, like insurance ID, name, or address. Furthermore, the patients can choose which doctor they would like to send the form to. This is important since many medical offices are shared by multiple doctors. For our concept, it is necessary to specify which doctor is the recipient of the form in order to initiate the sTeaLs protocol with the corresponding doctor. The second part explains the functionality of the form to the users and the implications of the layered-encryption. We chose not to place this text at the top but over the part that it is relevant for. This choice was made due to the design principle of proximity. The third part consists of the two text areas that define which information is considered sensitive and will be encrypted using the sTeaLS key and which information is transmitted via regular TLS. The only choice that is available in this version is the choice of which information is put in which text area. This is the control group to investigate the control paradox.

The second view is designed for the doctor: The structure is the same as the online form, but the first part with the patient's information is not editable and the second text area is highlighted in red as a warning for the doctor. This serves two reasons: first, the doctor is reminded that the data contained in this text area is meant to be treated with care since it is considered sensitive by the patient. Secondly, due to the additional encryption, the firewall does not scan the content. In our scenario, we assume that cross-site scripting is prevented by technical means, but there remains the risk that social engineering, such as phishing, is used by a malicious user. The red highlighting should alert the doctors to be careful when interacting with this data (e.g., clicking on links in this text area). Additionally, a warning text is displayed over the text area: 'Content not checked. Handle with care'. A possible phishing attack could be the following text: 'I got injured and took a photo of the wound. I attached the picture with the following link <URL>. Please tell me if I should go to the hospital with it.'

The third view resembles the answer email from the doctor to the patient. It also resembles the structure of the online form but without the information text. Furthermore, the partial encryption feature is simulated: the second text area is hidden behind a button with 'Decrypt and show', which is hidden when clicked. This simulated the following aspect of the concept: The second text area (containing the sensitive data) is removed from the email if the user is not logged in to the key wallet. If the user is authenticated, a 'decrypt and show' button is displayed instead of the text area. If the user clicks on it, the sensitive data is decrypted temporarily and displayed. This feature ensures that the data marked as sensitive is never stored in plain text on the patient's computer.

Although this demonstrator is just a mock-up, we implemented some quality-of-life features: The input fields of the three views are interconnected so that the user's inputs are also put in the corresponding input fields of the other views. In the case of the second view, the doctor's answers were updated only in the third view (answer email) and not in the first view (online form). Furthermore, to increase accessibility, users can use the Tab-key to switch between the input fields in the correct order.

(2) The second version allows the users to specify which part of the form should be sent via the stealth channel. Each input field has a corresponding checkbox that marks this field as confidential. Therefore, the users can fully customize their level of protection. This is the experimental version since it comes back to the problem of our research gap: Which trade-off between confidentiality and usability (possibility of keyword search) do users make? This version is referred to as the *customizable version* (see Fig. 11.5).

The customizable version is quite similar to the non-customizable version, but does not have the second text area for the sensitive information. Instead, each input field has a corresponding check box to mark it as sensitive. The information text is at the top to introduce the privacy mechanism first and explain the implications of layered encryption. For further convenience, we added an option to receive the requested receipt or referral letter via mail. Other quality-of-life features were a mark-all option for the check-boxes, and we simulated an adaptive form by making it responsive based on the choice of the users' request. In the doctor's view, all input fields marked as sensitive are highlighted in red. Furthermore, an information text indicates that the highlighted information was not checked by the firewall. The third view resembles the answer email. Similar to the non-customizable version, the input fields marked as sensitive are not visible if the user is not authenticated. Our demonstrator covers the case that the user is authenticated and hides the sensitive (encrypted) information behind a button ('decrypt and show'). After clicking it, the button becomes invisible, simulating the process of temporarily decrypting the data and rendering it into the form. Because of the limited space in the input fields for the house number and postcode, the button just states 'show'.

In both versions, we did not make any required fields since we wanted all data to be considered equally and avoid suggestive biases. This way, we can examine which data is considered sensitive by the participants of our study. During the interviews, we used questionnaires on a self-hosted instance of LimeSurvey[4] to get demographic information and some quantitative feedback. In order to test the usability of the demonstrators, we used the short version of the user experience questionnaire (UEQ-S)[5] for both demonstrators.

11.5.1.3 Procedure

Each session consisted of several parts which were completed by a single participant in about 30 to 50 minutes. The interview guideline can be found in the Electronic Supplementary Material A.4.3.

First, we had a briefing, and the participant signed the informed consent form. The form was sent to the participants prior to the study in order to allow them to read it without the context of the study setting, which would make it harder for them to refuse participation. After studying the hard copy of the informed consent form and signing it, the first questionnaire was filled out to collect demographic information about the participants, investigating their privacy knowledge using the Online Privacy Literacy Scale (OPLIS)[6]. Since OPLIS provides a norm table for the whole population, we decided to classify the participants into three groups, i.e., low,

[4] https://www.limesurvey.org/

[5] https://www.ueq-online.org/

[6] https://oplis.de/index_eng.html

medium, and high privacy literacy. The thresholds were set to divide the population so that each group represents around a third of the demographic spectrum, according to OPLIS. In our sample, seven participants were classified as having low privacy literacy, 13 as having medium privacy literacy, and 22 were classified as having high privacy literacy. This means that we have a bias towards users with a higher privacy literacy. Nevertheless, as described in Sect. 11.5.3, privacy literacy does not have an impact on the version of our demonstrator which is preferred by the participants. Additionally, we included the items from the quantitative study for comparison.

The second part of the study was the hands-on experiment. Since we had two demonstrators, this part consisted of two rounds: We alternated the order in which the participants encountered the two versions with each session to avoid sequence effects: Participants with an odd participant-number (P1, P3, P5, etc.) encountered the non-customizable version first and the customizable version in the second round. Participants with an even participant-number (P2, P4, P6, etc.) encountered the customizable version first and the non-customizable-version second. Every participant was asked to interact with both demonstrators and fill them out in a way that would make sense to them. This means that the participants were not required to fill out the form using their data, such as their real addresses, but fill in data that was plausible to them in such a scenario since we were investigating the three different views for each demonstrator and the data should be recognized in the other views. The scenario was to order a receipt for medicine or a letter of referral to another doctor via the online form. After filling out the forms and checking the views for the doctor and the response email, the UEQ-S questionnaire was filled out, and the process was repeated for the other demonstrator.

After finishing the tasks with the demonstrators, the participants filled out a questionnaire which compared both versions. The participants had to choose which version they preferred and explain their choice in a few words. Furthermore, users should rate whether they would use such a tool in real life on a 5-point-Likert-scale (Would definitively use—Would definitively not use). During the evaluation, we used the Think-Aloud-Method [337] and recorded the audio.

As a final part, we had a discussion with the participants. For this, we prepared guiding questions which can be found in the Electronic Supplementary Material A.4.3, investigating aspects they (dis-)liked and reflecting on their behavior during the experiment.

11.5.2 Data Analysis

In our data analysis, we followed the first six steps of the model of Olson, McAllister, Grinnell, Gehrke Walters, and Appunn [338] for applying the Constant Comparative Method (CCM), a special form of grounded theory. The authors claim that their improved model increases the credibility of CCM. Furthermore, we chose this method because it allows for iterative development. The audio data was transcribed, and the questionnaires were converted into tables. After the (1) initial sighting of the material, interesting quotes and important topics were gathered and clustered. Based on this, (2) a codebook was developed, which can be found in Electronic Supplementary Material A.4.6. Utilizing this codebook, two researchers analyzed the material independently (3), marking which participant contributed to which code. One researcher listened to the audio recordings and analyzed the open questions from the questionnaire, comparing both demonstrators, while the other read transcripts of the audio recordings. Comparing the coding (4) showed that for 40% of the codes, the coders had identical results; for the other 60% of the codes, the error rate varied between eight and 82% (mean: 18%), indicating that participants brought up further topics in the questionnaire. Afterward, inconsistencies were discussed (5), specifying the code book's criteria for applying codes (6). Analyzing the coded data, we found that after participant P34, no new topics were brought up, and we assumed that saturation had been reached [339]. The analysis was done in German since most interviews were conducted in German (one participant preferred to answer in English). The finished results and the quotes for the result section were then translated into English as literally as possible. Afterward, some phrases were polished during proofreading to avoid confusion.

11.5.3 Results

11.5.3.1 Preferred Version

After the participants tried both versions of the demonstrator, they were asked which version they preferred. 31 participants stated that they preferred the customizable version, and 11 preferred the non-customizable version. Since we alternated the versions so that one-half of the participants started with the customizable version and the other half started with the non-customizable version, we examined whether a sequence effect occurred. We can deny this because 71.43% of those who started with the customizable version liked it most (15 of 21), and 76.19% of those who started with the non-customizable version also liked the customizable version most (16 of 21). We also examined whether privacy literacy influenced the participants' preferences. We can also deny this bias: Of the participants with low privacy literacy, 71% preferred the customizable version; of the participants with medium privacy literacy, 69% preferred the customizable version; and of the participants with high privacy literacy, 77% preferred the customizable version.

The participants stated a variety of reasons why they preferred one version over the other: One participant who preferred the non-customizable version stated, that it could confuse the user to decide for every input field whether it should be only visible for the doctor, which leads to an overhead that costs the user more time (P33). P22 stated that the customizable version could lead to miscommunication. This is a concern which was shared by P27, who stated the possibility to 'miss-click': "The checkboxes are intuitive, but you could confuse the rows and check the wrong box." Furthermore, the non-customizable version would provide a more strict separation between the sensitive and the non-sensitive data. This was also pointed out by P11: "You notice that the fields are clearly separated. This gives me the feeling that it is secure. For people who are not familiar with technology such a feeling could be more important than the technology itself." The possibility of usage errors was pointed out by P8, who preferred the non-customizable version because it was "easy to use", a sentiment that was shared by four other participants (P13, P17, P28, and P30). Additionally, P31 stated that the customizable version would be inefficient: "It is inconvenient that you have to click every single thing". P17 further stated that the information that could be marked sensitive in the customizable version would be relevant for other staff members and not only for the doctor. This was a point of criticism that other participants shared.P2 pointed out that the personal information was needed to identify them more easily, and the only relevant information worth protecting would be the reason why the patient needed the doctor's help. The non-customizable version would offer the possibility to do so. P3 argued similarly that staff members need information such as the address to support the doctor efficiently. If everything could be hidden from them, efficiency would suffer. Furthermore, independently of their preference, P11, P14, P19, P21, P24, P25, and P28 pointed out that hiding data from staff members would lead to administrative problems. The efficiency of the work processes was also a concern for those who preferred the customizable version, which affected their choice during the interaction with the demonstrators (see Sect. 11.5.3.3).

The participants who preferred the customizable version did so because of usability, perceived security, and the ability to customize and control their interaction with the doctor. For example, P20 described the customizable version as "easy and clear. The ability to select single elements works fast and is intuitive". P20 continued with the complaint that the two different text areas were cumbersome since the user had to write two texts and was prone to reveal information without intending to do so if the concept was misunderstood. That the two inputs from the non-customizable version are cumbersome was shared by P5, P9, and P38, who described the customizable version as less "redundant". P21 proposed an approach with two text areas for the patient's message but for another use case: The form should provide the possibility to write a message to the doctor and a second one to the other staff to enhance the performance for each request. P14 reported that the customizable version had better usability. P14 praised the better guidance of the user and that it was intuitive, which information was considered sensitive, a fact that was also brought up by P37. P42 added that the fact that the checkboxes allowed a decision close to the corresponding information was a great benefit. The ability to protect all personal data was important to P40 and P29, who stated an increased feeling of security. P13 added: "It's all health information, it's all sensitive in some way." P26 had the impression that it was less likely to unintentionally reveal data in the customizable version compared to the non-customizable version due to the fine-granular customization. P36 added that this ability to encrypt everything enhances the feeling of privacy. Overall, empowering users to decide the level of protection for their data was the most important factor as to why participants preferred the customizable version. In particular, P6 had the feeling of having more freedom to choose over his data and further stated that "all data that I do not encrypt, I chose not to encrypt". This was confirmed by P15, stating, "You can decide yourself which data should be protected". P18 also favored this choice: "I think I like the [customizable version], because at least there I had a choice which information I want to share". This self-controlled disclosure of data was also perceived as part of sovereignty over one's data. P24 expressed a similar aspect: By being able to customize the visibility of her data she can match her personal preference of what is considered sensitive. This was confirmed by P41, who stated that "the decision is taken away" from her in the non-customizable version. The trade-off between privacy and usability was brought up by P19: "In this [customizable version] I have more options to define what should be treated confidentially. This leads to a slight overhead (just some clicks), which is not a hindrance for me."

11.5.3.2 Usability of the Demonstrators

We used the UEQ-S to evaluate the usability of our two demonstrators. With this measure, we wanted to verify that lacking usability had no influence on our qualitative results. The usability scores in Electronic Supplementary Material A.4.5 attest to excellent usability according to UEQ-S (values higher than 4.5) and show that the demonstrators were designed well enough not to pose a bias to the investigation of the privacy paradox. This is supported by statements from the participants as well: For example, P18 said: "I think I like the [customizable version] because at least there I had a choice which information I want to share. Other than that, I think both are almost the same". In general, the feedback of the participants focused on the two different options to apply layered encryption and protect their data. This may be due to the fact that we transparently communicated that the demonstrators were mock-ups or that the look and feel of the Axure assets emphasized that it is not ready to use. Feedback on the rest of the forms was limited to aspects like the wish for colored forms or a date picker to specify the date of birth. The latter was not included in the demonstrators because no such option was available in the mock-up tool.

11.5.3.3 Which data is considered sensitive?

An overview of how the participants selected the data can be found in Table 11.1. Please note that 5 participants were removed from this overview because they stated that they did not set the check-boxes like they would in real life but randomly to see the effects on the other views. They still provided valuable feedback on the demonstrators and the use case, which is why they were only excluded from this analysis.

The information that was considered sensitive by most of the participants was the actual medicine or the type of doctor they wanted to be referred to. Interestingly, these were exactly those fields that correspond to the text areas in the non-customizable version. However, despite the seeming indifference between the two versions with this configuration, the participants preferred the sovereignty over their data in the customizable version (see 11.5.3.1). Some participants considered the town they live in as less sensitive than the street they live in since this information could be used more effectively to find them in real life. During the evaluation of the demonstrator, many participants stated that they suspected that some of their personal data would be necessary to process their request in the doctor's office. Therefore, many of them did not protect their personal information, fearing that their request might not be processed due to missing data. This is, in part, the fault of our design choices: We did not want to make 'required fields' in the form because we wanted to investigate which data is considered sensitive. Highlighting some of it as required would be a bias.

The choice of which data should be considered sensitive was also addressed in the discussion. For example, P12 reported that a female friend voiced her last name and date of birth in a medical office, which is a common practice at the registration desk. Someone who happened to be there too, was able to find her on social media with this little data. P12 was afraid that someone with more resources could do even more than that. Another concern was raised by P15: By disclosing personal data, it becomes more likely that someone could attempt identity theft to gain even more private data.

Table 11.1 Protected fields in the customizable version (•) vs. the unprotected fields (empty)

Field	P1	P3	P5	P6	P7	P8	P9	P10	P12	P13	P14	P15	P16	P18	P19	P20	P21	P22	P23
Insurance Number	•						•								•	•	•		
Last Name	•						•		•			•				•			
First Name	•						•		•			•				•	•		
Street and House Number	•				•		•		•			•		•		•			
Town and Postcode	•				•		•							•		•			
Date of Birth	•				•		•		•					•		•	•		•
Type of Request	•			•		•	•			•	•		•		•		•		
Receipt or Referral	•	•		•		•	•			•	•		•	•	•	•	•		•
Comment	•		•	•	•	•	•	•	•	•			•	•	•	•	•		•
Field	P24	P25	P26	P27	P28	P29	P30	P31	P32	P34	P35	P36	P37	P38	P39	P40	P41	P42	
Insurance Number	•	•	•	•				•		•		•		•					
Last Name			•					•		•		•						•	
First Name			•					•		•		•		•			•	•	
Street and House Number	•		•					•		•		•		•		•		•	
Town and Postcode	•		•					•		•		•						•	
Date of Birth			•					•		•		•		•			•	•	
Type of Request	•		•	•				•	•		•	•						•	
Receipt or Referral	•	•	•	•		•		•	•	•	•	•	•				•	•	
Comment	•	•	•	•	•			•	•	•	•	•					•	•	

P26 stated that he would provide more information (by not selecting the checkbox) if he knew which information was used for which purpose and therefore, could evaluate how he can support the processing of his request by revealing information without risking the negative effects of the disclosure. P34 was also concerned about the extra work for the staff: He would always encrypt the data because it would not make sense to him otherwise: If non-sensitive information is not encrypted, that means that all encrypted data is sensitive data, making the communication suspicious. He assumed that many people would use the encryption if available, creating more work for doctors, which could become a problem in Germany. This assumption is confirmed by P36 who stated that she would always encrypt every possible information if this did not increase the time for processing her request too much. Other participants also stated the trade-off between more disclosure and faster processing of their requests (P18, P32, P42). Another fear was revealed by P21 and P35: they were afraid that their request could not be processed properly due to missing data.

Another interesting aspect was brought up by P13, who stated that personal experience would highly impact the choice of protected data and that people who experienced negative consequences of lacking privacy would act more sensible in this regard. P14 also mentioned the connection between personal experience with privacy and behavior in relation to privacy: "A person who never bothered with data protection or their personal data might get incentivized [by the customization option] to think about what data should be protected. This might raise awareness in society".

For further insights into the consistency of privacy behavior, we correlated the answers from the questionnaire ("Which data would you disclose to third parties without concern?") to the choices in the demonstrator for the four items first name, last name, address, and insurance number (the latter correlated to the item "Customer ID"). We found that seven participants encrypted their first name despite claiming to disclose it ("over-encrypted"), and two participants did not encrypt it despite claiming not to disclose it in the questionnaire ("under-encrypted"), leading to a total discrepancy of 24%. Four participants "over-encrypted" and 14 "under-encrypted" their last name (discrepancy 48%). In the context of their address, one "over-encrypted" and 22 "under-encrypted" (discrepancy 62%). Finally, six participants "over-encrypted" and four "under-encrypted" their insurance numbers (discrepancy 27%). These discrepancies might result from the use case of our evaluation: In the pre-study, we did not provide a special context on purpose, but in the evaluation of the demonstrators the participants were confronted with the context of a medical office. We confirmed this by reviewing the audio recordings from the experiment: While filling out the forms, the participants did not only consider their personal preferences, but the workflow of a medical office. Therefore, they hesitated to encrypt information that they thought was necessary to identify them as a patient and to process the request (P19, P28, P35, P42). P6 considered the encryption as a measure to direct the data to different addressees: "Information such as addresses is not interesting for the doctor, but the assistants". Similarly, P2, P5, P21, P25, and P34 did not encrypt some of the data to minimize the effort for the doctor and the assistants. Not having all available information or forcing the doctors to do administrative tasks like sending a letter themselves was not deemed justifiable in that use case (P25. P28. P34). This explains the over- and under-encryption from our correlation: Last names are under-encrypted to identify the patient. In Germany, when entering a medical office, patients state their names and "verify" their identity by stating their date of birth. This is not a security feature but is intended to ensure that the correct patient data is used and that mistakes due to common names are avoided (e.g., identify the correct John Smith). The first names are not so essential for this process since the last name and birth date are often sufficient for identification. Thus, the over-encryption can be explained. In some cases, the participants thought it would be sufficient to leave the insurance number unencrypted to identify them, resulting in under-encryption. The under-encryption of the addresses also originated from the use case: In order to send receipts, the staff members need to know the addresses. Therefore, many participants did not encrypt them on purpose to support the workflows of staff members.

11.5.3.4 Form-based Communication

In our study, we found that form-based communication is a promising approach. After the interaction with the demonstrators, the users were asked in the questionnaire if they would use such a form if it would be available in real life. The possible answer ranges were: 'definitely', 'likely yes', 'I don't know', 'likely no', 'definitely not'. Two participants voted for 'I don't know', while 13 voted for 'Likely yes' and 27 voted 'definitely'. Although this result may be affected by the acquiescence bias, we confirmed that the possibility of communicating via an online form was highly appreciated.

Reasons for that are not only our novel concept for partial encryption of sensitive data but also the ability to communicate with doctors more easily, reducing barriers to seeking help. The latter was confirmed by participants who stated that it was easier to formulate embarrassing or stigmatized topics in text than in a conversation in person or via telephone (e.g. P3, P5, and P27). Furthermore, P4, P12, P20, and P42 complained about the ability of bystanders to eavesdrop on telephone calls with the medical staff. In Germany, the calls are answered by staff at the registration desk, which makes it possible for other patients waiting in line to listen to the conversation.

P20 stated: "It is uncomfortable when the staff talks loudly about you and others in the medical office can hear it. Even if it is nothing sensitive, like going to the dentist". In general, communication via telephone is perceived as cumbersome to some (P6, P13, P16, P31, P37) or even as reason for discomfort or fear to others (P4, P6, P10, P19, P20, P26, P30 P32). The more flexible way of communication via an online form was pointed out by twelve participants. Users do not have to wait (very long) for their calls to be answered or can send requests whenever they like without having to pay attention to the opening hours of the doctor's office, which is a major problem for people who have to work shifts that collide with those time slots. For example, P28 stated: "I am a student, and I am at the university all day. I can't go out and try calling the medical office a thousand times a day. An email or a form is easily done. Therefore, it makes a lot of sense [to use this way of communication]". Besides perceiving communication via telephone as cumbersome, eight participants mentioned telephone anxiety, which affected them or people they knew (P4, P6, P13, P16, P19, P20, P26, P30). On the other hand, some participants mentioned that they would still use the telephone for urgent cases (P9, P40).

11.5.3.5 Use Cases

We did not comment on the use cases mentioned by the participants or discuss whether they are hypothetical or based on their own experience since many of the use cases are very sensitive (e.g., depression or domestic violence), and we did not want to trigger traumas or embarrass the participants. Furthermore, we refrain from using the participant ID in this section to provide further anonymization. Instead, we will include just the number of participants who mentioned a certain topic for each topic.

We distinguished the use cases mentioned by the participants in two scenarios: First, we investigated for which use case participants consider it important that the answer email is partially encrypted so that people in their surroundings cannot see the data they marked as sensible, even in the case of "shoulder surfing":

One of the most mentioned reasons for this scenario was getting a severe or lethal diagnosis and trying to comprehend and process this information by themselves before anyone else finds out. Furthermore, the participants wanted to decide for themselves when (or whether) the diagnosis would be disclosed to others (12). One category in this section can be summarized as "housing situation": Participants reported many use cases where it is important to have strong privacy features at home. The reasons ranged from "mistrust at home" (4) over living in shared flats (8) to having a divorce but still living together (1). Four participants even mentioned domestic violence. Regarding the housing situation, participants mentioned the parent-child relationship: Three participants did not want to disclose the parents' diagnoses to their children to prevent them from worrying. On the other hand, children should be protected from parental surveillance (8). Furthermore, three participants mentioned the issue of females wanting to go to the gynecologist against their families' will. Another category is about intimate topics: Eleven participants stated intimate topics, in general, were confidential. Others were more precise and specifically addressed the topics of (unwanted) Pregnancy (6), abortions (3), or results from the urologist (6). Similarly, taboos or topics stigmatized by society were mentioned: Besides the general topic (13), the participants mentioned Sexual Transmittable Infections (STIs) (12), mental problems (e.g., depression) (9), drug abuse including alcohol (3), and help for suicidal people (1). Furthermore, the fear of discrimination was brought up: One participant said that any information that could prevent him from living his life as before if it were to be leaked was worth protecting. Others reported more specific reasons for discrimination: LGBTQ topics (2) or (mental) disablement (1). Further use cases were the handling of legal documents (21) or communication with schools (3), as well as religious reasons (3). For example, women might not want to go to a male doctor due to religious beliefs, and professions like gynecology may be stigmatized by some beliefs.

Secondly, we investigated use cases for privacy towards other people in the doctor's office, achieved by the additional encryption: The reasons for privacy within the medical office are divided into two groups: The actual topic, why someone attends the medical office and social aspects. The topics that were considered most sensitive were intimate issues (8), mental problems (19), and embarrassing or stigmatized problems (17). The social aspects mostly cover the issue of rumors (3). For example, one participant stated that her grandfather did not want to go to the doctor because he feared potential rumors. Furthermore, having relatives or family members among the medical office's personnel was a concern for 19 participants. Eight participants also mentioned that living in a rural area is problematic: There are fewer doctors, and most people know each other, which makes it impossible to go to the doctor without anyone knowing it. A corresponding quote from a participant who lives in the city was: "If I know someone in the medical office and I do not trust them, I go to a different medical office".

On the other hand, participants explicitly stated that they have a general trust in medical staff (10). Reasons for this differed: Some personally trust the staff, others rely on the laws and regulations in Germany which enforce secrecy for medical data. Regarding their level of trust, some participants also differentiated between the doctor and the assistants. Furthermore, one participant even differentiated between the different staff members: She mentioned that at her gynecologist, there is always an assistant present. Therefore, she trusts those assistants for treatment more than those who are in charge of administrative tasks. A limitation of this system was mentioned by one participant who stated that frequently changing personnel among the staff would reduce trust.

11.5.4 Limitations

In our qualitative evaluation, we face the limitation that our sample consists of relatively young and well-educated participants. This makes our results less representative. To evaluate to which extent this bias affects our results, we compared our sample with the representative sample from the pre-study. For this reason, we included the questions from the quantitative survey in the questionnaires of our qualitative evaluation. By statistically assessing the difference between our qualitative sample and the representative sample, we can estimate to which degree the results can be generalized. We compared the following samples: *Sample A*: answers from the participants of our qualitative study (N=42); *Sample B1*: answers from the age group of 18-44 from the representative survey (N=378); *Sample C1*: answers from the representative survey that are not part of B1 (N=633); *Sample B2*: answers from the representative survey from the age group of 18-44 with a similar education level compared to sample A (N=222); *Sample C2*: answers from the representative survey that are not part of B2 (N=785). In samples B2 and C2, we exclude the answers of those who stated "other" when asked for their highest educational degree. We conducted a Mann-Whitney-U-Test to compare two samples at a time. For each item from each question we compared sample A to B1 and B2 in order to evaluate how the answers of our qualitative sample differ from the corresponding group in the representative sample. Additionally, we conducted the test for samples B1 vs. C1 and B2 vs. C2. This way, we could evaluate whether age (or age and educational level) had an impact on the answers to our questions. Furthermore, the sums of the answers were compared as well. This way, a Pearsons correlation efficient was determined as the effect size [340] and interpreted according to Cohen [341]. We specified p=.05 as the significance level ($|r| < 0.1$ = very weak; $0.1 < |r| < 0.3$ = weak; $0.3 < |r| < 0.5$ moderate; $|r| >= 0.5$ = strong). For the first question, all differences in the Mann-Whitney-U-Test were weak, very weak, or insignificant, with one exception: the difference between sample A and B2 was moderate for the item "chaplain".

Furthermore, all effect sizes were negative, which means that A-samples indicated a higher importance than B-samples, which in turn indicated a higher importance than C-samples (higher because in the Rep study '1' stood for "very important" and '5' for "very unimportant"). For the second question, all differences were weak, very weak, or insignificant. Again, all (significant) effects were negative, which means that A-samples were more willing to share data than B-samples, which in turn were more willing than C-samples. However, if we sum up the answers to the questions on sharing data, we see that there is a moderate effect, which means that the A-sample was more willing to share data than the B-sample. To provide full transparency, we added a detailed table for all items in the Electronic Supplementary Material A.4.1. To summarize, we have limitations regarding the age and the education of the participants in the qualitative sample, but the differences do not render our qualitative results useless.

Another limitation is that some of our findings are not exclusively related to our concept. For example, providing an alternative to the telephone for communication with doctors can be achieved without the additional protection features. However, those findings are relevant for our paper since we propose a concept that increases privacy and security without the need for a trade-off. Therefore, the additional findings provide insights necessary for potential adoption into practice. Further limitations are the qualitative nature of our evaluation and the focus on the German population. Studies in different countries may yield different results (see Sect. 11.6.4).

11.6 Discussion and Conclusion

In this paper, we were able to shed the first light on our interaction concept D^3. In the following, we will discuss topics that were brought up during the investigation of suitable use cases and evaluate our concept with the first iteration of demonstrators.

11.6.1 Evaluating the Control Paradox

In our study, we found that the majority of our participants preferred the customizable version. Since this was independent of the sequence they encountered the demonstrators and the privacy literacy of the participants, we conclude that the ability to customize their own privacy is important to users, confirming these claims in the literature [41]. Some participants reported personal experiences with negative consequences of data leakage and were eager to encrypt all data, while others, who did not perceive the necessity of encrypting personal data still liked the possibility of doing so. Besides the fact that many customized the level of protection similar to the level of protection of the non-customizable version, the circumstances were different: The unprotected data is the same for both versions, but one version forces the disclosure of data to staff members and the other leaves the decision to the patient.

On the other hand, the statistical comparison to the representative sample showed that besides no significant difference in the answers of the questionnaire items, there was a tendency to disclose more data among our qualitative sample while claiming a higher importance of protection. This may be due to the sample size or a confirmation of the control paradox stated by Brandimarte, Acquisti, and Loewenstein [34] and Boer, Herder, and Lieshoud [35]. We can rule out the influence of the demonstrator design because the questions used for comparison were answered before the interaction with the demonstrators. In summary, the qualitative sample answered the questionnaire items similar to the representative sample, with a slight tendency to rate the protection of communication higher (item 1: "How important is it to you to secure communications with the following people and institutions?") and a slight tendency to protect less data assets (item 2: "Which data would you disclose to third parties without concern?"). Therefore, we can assume that the settings of the checkboxes would be similar for a larger sample. Considering the non-significant derivation, a more representative sample would possibly check some boxes more.

One of our participants gave a possible explanation for the control paradox: By granting the users the freedom to choose the level of protection they want, the disclosed data is voluntarily given away. Denying the users this choice makes them feel like the data is taken against their will, which results in the feeling that the company is the adversary. Withholding as much information as possible from them is like fighting back. This could be an explanation of the control paradox: If users have the power to decide, they can evaluate, based on individual information, whether it may be disclosed or not, which could result in them revealing more data due to a better feeling. However, if they are forced to give away data, they will try to withhold as much as possible.

11.6.2 Practical Relevance

In our study, we found several implications for the real-world adoption of our interaction concept. First, we confirmed that one barrier would be the potential increase in workload for doctors. This was the reason why we designed the concept to be voluntary. Forcing someone to use it who will circumvent the purpose of it due to lacking capacities would be fatal. Therefore, we recommend our system for doctors who have the capacity for an additional workload or those who wish to address a broader field of patients. Another target group for our concept are counseling agencies for sensitive use cases. Our participants reported some very critical use cases where it is crucial that patients can communicate securely and secretly. Examples of this are use cases in which people face repression from their surroundings, such as religious reasons, repressive families, or even domestic violence. Social workers who provide support in these situations could benefit from our system since it is suitable for covered communication due to the use of the sTeaLS protocol. Due to the fact that it is not distinguishable from the outside if a regular TLS handshake was conducted or the sTeaLS protocol was used, it is deniable that the second key exists. Therefore, steganographic methods can be applied to deny that a second key ever existed to adversaries who can monitor the network traffic.

Introducing our concept to the public would confront those people who are not concerned about privacy with a potentially undesired mechanism. However, our concept benefits all users, not only those who value privacy. For example, the use of form-based communication was perceived as a major improvement to the current situation. Due to lacking digitalization, the telephone is the only viable option for many patients. This perception of lacking digitalization is not limited to our sample, as a recent investigation shows [342]. The preference for the use of online forms was stated by many participants of our study. We presented research in the related work section that emphasizes the role of trade-offs between security and usability [42–44]. However, with our interaction concept D^3, we propose a solution that increases security while being more usable than the current status quo. This way, we tackle the challenge of the trade-off between confidentiality and usability by improving the usability of communication with doctors compared to the current situation while adding the option for additional encryption. Therefore, we do not need compromises but offer better usability with better confidentiality. Based on the enthusiasm of the participants for real-world adoption of our concept, we are confident that this technology can improve the lives of many people, especially of those who perceive the current best practice as a hindrance to getting help. By providing online forms as an alternative, people who cannot use the telephone due to their physical condition, time constraints, or telephone anxiety can easily contact a medical office while being able to customize the level of confidentiality. Therefore, benefits are provided even for those who do not value privacy-preserving communication.

11.6.3 Impact on the Iterative Design of the Interaction Concept

The qualitative study was the first step in the evaluation of our concept. Valuable feedback from the participants will be used to enhance the interaction concept even further. Many participants mentioned that they did not know which information was necessary for the rest of the staff to process the request. In the study setting, we abstained from using ‘required fields’ to avoid a bias in the selection of ‘sensitive data’ by the participants. In a real-world scenario, each medical office or company that would use our concept would have individual forms for their services suited for their work processes.

‘Express forms’ could be offered to address the concern of some participants that the partially encrypted data would lead to slower processing because only the doctors can see and process the data. By clearly informing the patient that those forms do not provide the partial encryption feature, the patients can decide whether faster processing is worth disclosing their data.

11.6.4 Transferability of Results

Our research focused on the German population. To evaluate whether our results are applicable to other countries, we investigated existing literature: Firstly, in studies comparing multiple countries, Germany belongs to the countries with higher privacy and security concerns. Herbert et al. [343] investigate the misconceptions regarding security and privacy topics (12,351 participants in 12 countries: China, Germany, Great Britain, India, Israel, Italy, Mexico, Poland, Saudi Arabia, Sweden, the USA, and South Africa). In their survey, they found differences regarding security and privacy misconceptions between Western and non-Western countries. Germany, therefore, differs from non-Western countries but is similar to other Western countries. Prince et al. [344] investigate the level of privacy concerns of different European countries by surveying online privacy literacy. Using the Eurostat 443 dataset (n=26,526; 28 countries), they investigated the privacy concerns regarding access to their data, monitoring, and confidentiality. Compared to the reference country France, Germany belonged to the group of countries with higher privacy concerns (Ireland, United Kingdom, Denmark, Malta). On the contrary, there was also a group with fewer privacy concerns: Cyprus, Czech Republic, Estonia, Hungary, Latvia, Lithuania, Poland, Slovakia, Bulgaria, and Croatia. Harbach et al. [345] investigated the smartphone locking behavior in eight countries (n = 8,286): Australia, Canada, Germany, Italy, Japan, Netherlands, United Kingdom, and the United States. The authors found that data protection was rated higher in Germany and Japan than in the US sample. Utz et al. [346] investigated the acceptance of contact tracing apps during the Covid-19 pandemic in Germany, USA, and China. The German and the US sample were hesitant towards contract tracing, while the Chinese sample was fonder of this concept. The authors assume that the familiarity with similar systems in China reduced the barriers for adoption. On the contrary, the German and US participants had greater concerns regarding governmental surveillance. Secondly, we looked into studies that make a direct comparison between Germany and another country: Ilhan and Fietkiewicz [347] investigate the difference between German and US users regarding their knowledge and behavior towards activity tracking. Surprisingly, the US sample was more likely to deactivate accounts or request the deletion of data than the German sample, although the German sample knew more often about the possibility of deleting the data. Further analysis showed that US users were more likely to inform themselves about the company, their reputations, and the Terms of Service, while the German sample was more likely to read the Data Privacy Policy. However, both samples had a similar rating of data sensitivity and concerns about data misuse. Pleger et al. [348] conducted a media analysis and survey with 1000 participants from Germany and the UK (500 from each country). While both populations state the importance of data protection and data security as well as have similar trust in the government, the UK citizens had a greater concern about disclosing personal data on the internet than the Germans (7.34 out of 10 vs. 7.09 out of 10). Similarly, the concerns for 'fraudulent use of data', 'data theft', 'identity theft', and 'electronic manipulations of elections' were slightly higher in the UK compared to Germany.

Non-Western societies, in particular, are different in their perception of security and privacy topics. On the other hand, Germany was relatively comparable to the UK and the US. Besides some differences of opinion on certain topics, all three countries are among those who value privacy and security highly. Another finding from the literature is the role of the state in the perception of societies. Reuter, Kaufhold, Schmid, Spielhofer, and Hahne [271] show that Germans trust the state and hold it responsible for emergency management, while this perception differs in other European countries. Therefore, the findings from our study are relevant to other countries as well. Nevertheless, before adopting our proposed interaction concept in other countries, further regional research is recommended.

11.6.5 Future Work

With this study, we confirmed that our interaction concept can address relevant use cases and provide benefits for civilians. After discovering several use cases (see Sect. 11.5.3.5), a possible next step would be to conduct expert interviews with professions that are affected by these use cases and could benefit from our system. Furthermore, to allow easy adoption into practice, a form generator should be implemented to allow for an easy design of individual forms for each use case. Additionally, the backend implementation to host those forms should be made public (e.g. open source).

11.6.6 Conclusion

In this paper, we presented and evaluated an interaction concept to bring state-of-the-art cryptography into usage in everyday scenarios. By utilizing the sTeaLS protocol [284], users are provided with the possibility to add an additional layer of encryption for their data, which also provides protection from middleboxes. In this paper, we propose an interaction concept (D^3) that utilizes this protocol for communication with specific recipients within a business or an organization, leaving the decision to the users which part of the message is additionally encrypted. To find the most relevant use cases for this partially-accessible encryption, we conducted a quantitative survey with 1011 participants representative of the German population for the criteria of age, gender, education, and income. After identifying the interaction with doctors as the most suitable use case, we investigated whether new privacy solutions for this use case are appreciated by end users, and further development in this area should be continued by conducting a qualitative evaluation with 42 participants. Enabling users to decide over their data by themselves and its protection is prone to the privacy paradox: The tendency to disclose more data if more control over it is granted. To investigate this, we decided to utilize two demonstrators. One with no customization features that offer a basic functionality and a version that allows a full customization of which part of the message is additionally encrypted. The results of our study show that the possibility of partial encryption is highly appreciated and could be beneficial for several critical use cases. One major theme among the participants was the sovereignty over their data. Even if they disclosed a lot of information, they appreciated that this was their own choice. This led to a preference for the customizable version over the non-customizable version in our study. Furthermore, besides the ability to utilize layered encryption, the possibility to communicate with doctors via an online form was perceived as an improvement to the current situation in Germany, where the telephone is often the only way to communicate with medical offices. Therefore, the proposed concept would reduce physical and social barriers to seeking help and combat the trade-off between confidentiality and usability by providing a custom level of protection while also offering better usability.

Acknowledgments This research work has been funded by the Deutsche Forschungsgemeinschaft (DFG, German Research Foundation)—SFB 1119 (CROSSING)—236615297 and by the German Federal Ministry of Education and Research and the Hessian Ministry of Higher Education, Research, Science and the Arts within their joint support of the National Research Center for Applied Cybersecurity ATHENE.

Bibliography

1. S. Linsner, F. Kuntke, E. Steinbrink, J. Franken, and C. Reuter, "The Role of Privacy in Digitalization – Analysing the German Farmers' Perspective," Proceedings on Privacy Enhancing Technologies (PoPETs), vol. 2021, no. 3, pp. 334–350, 2021. DOI: https://doi.org/10.2478/popets-2021-0050.
2. S. Linsner, F. Kuntke, G. M. Schmidbauer-Wolf, and C. Reuter, "Blockchain in der Landwirtschaft 4.0 - Empirische Studie zu Erwartungen von Landwirt_innen gegener dezentralen Services auf Basis von Distributed Ledger Technology," in Mensch und Computer - Tagungsband, F. Alt, A. Bulling, and T. Döring, Eds., Hamburg, Germany: ACM, 2019, pp. 103–113. DOI: https://doi.org/10.1145/3340764.3340799.
3. S. Linsner, E. Steinbrink, F. Kuntke, J. Franken, and C. Reuter, "Supporting Users in Data Disclosure Scenarios in Agriculture through Transparency," Behaviour & Information Technology (BIT), vol. 41, no. 10, pp. 2137–2159, 2022. DOI: https://doi.org/10.1080/0144929X.2022.2068070.
4. K. Demuth, S. Linsner, T. Biselli, M.-A. Kaufhold, and C. Reuter, "Support Personas: A Concept for Tailored Support of Users of Privacy-Enhancing Technologies," Proceedings on Privacy Enhancing Technologies (PoPETs), no. 4, pp. 797–817, 2024, ISSN: 2299-0984. DOI: https://doi.org/10.56553/popets-2024-0142. [Online]. Available: https://petsymposium.org/popets/2024/popets-2024-0142.pdf.
5. S. Linsner, K. Demuth, S. Surminski, L. Davi, and C. Reuter, "Building Trust in Remote Attestation Through Transparency – A Qualitative User Study on Observable Attestation," Behaviour & Information Technology (BIT), pp. 1–21, 2024, Publisher: Taylor & Francis, ISSN: 0144-929X. DOI: https://doi.org/10.1080/0144929X.2024.2374889. [Online]. Available: https://doi.org/10.1080/0144929X.2024.2374889.
6. S. Linsner, K. Demuth, M. Fischlin, and C. Reuter, "Decision-based Data Distribution (D^3: Enabling Users to Minimize Data Propagation in Privacy-sensitive Scenarios," Proceedings on Privacy Enhancing Technologies (PoPETs), no. 4, 2024, ISSN: 2299-0984. DOI: https://doi.org/10.56553/popets-2024-0113. [Online]. Available: https://petsymposium.org/popets/2024/popets-2024-0113.php.

S. Linsner, *Privacy Preserving Data Management*, Technology, Peace and Security I Technologie, Frieden und Sicherheit,
https://doi.org/10.1007/978-3-658-47445-4

7. G. R. Chandran, K. Demuth, K. Edalatnejad, S. Linsner, C. Reuter, and T. Schneider, "Encrypted MultiChannel Communication (EMC2): Johnny should use secret sharing," in 23. Workshop on Privacy in the Electronic Society (WPES'24), Salt Lake City, USA: ACM, 2024, pp. 34–39. DOI: https://doi.org/10.1145/3689943.3695051. [Online]. Available: https://dl.acm.org/doi/10.1145/3689943.3695051.
8. F. Kuntke, M.-A. Kaufhold, S. Linsner, and C. Reuter, "GeoBox: Design and Evaluation of a Tool for Resilient and Decentralized Data Management in Agriculture," Behaviour & Information Technology (BIT), vol. 43, no. 4, pp. 764–786, 2024, Publisher: Taylor & Francis, ISSN: 0144-929X. DOI: https://doi.org/10.1080/0144929X.2023.2185747.
9. S. Surminski, C. Niesler, S. Linsner, L. Davi, and C. Reuter, "SCAtt-man: Side-Channel-Based Remote Attestation for Embedded Devices that Users Understand," in Proceedings of the Thirteenth ACM Conference on Data and Application Security and Privacy, ser. CODASPY '23, New York, NY, USA: Association for Computing Machinery, 2023, pp. 225–236, ISBN: 9798400700675. DOI: https://doi.org/10.1145/3577923.3583652.
10. E. Steinbrink, T. Biselli, S. Linsner, F. Herbert, and C. Reuter, "Privacy Perception and Behaviour in Safety-Critical Environments," in Human Factors in Privacy Research, N. Gerber, A. Stöver, and K. Marky, Eds., Cham: Springer International Publishing, 2023, pp. 237–251, ISBN: 978-3-031-28643-8. [Online]. Available: https://doi.org/10.1007/978-3-031-28643-8_12.
11. F. Kuntke, S. Linsner, E. Steinbrink, J. Franken, and C. Reuter, "Resilience in Agriculture: Communication and Energy Infrastructure Dependencies of German Farmers," International Journal of Disaster Risk Science (IJDRS), vol. 13, no. 2, pp. 214–229, 2022, ISSN: 2192-6395. DOI: https://doi.org/10.1007/s13753-022-00404-7.
12. L. Buhleier, S. Linsner, E. Steinbrink, and C. Reuter, "Eine Klassifikation sicherheitskritischer UX-Design-Patterns," de, in Mensch und Computer - Workshopband, ser. Mensch und Computer 2022 - Workshopband, Darmstadt: Gesellschaft für Informatik, 2022. DOI: https://doi.org/10.18420/muc2022-mci-ws10-275.
13. F. Kuntke, V. Romanenko, S. Linsner, E. Steinbrink, and C. Reuter, "LoRaWAN Security Issues and Mitigation Options by the Example of Agricultural IoT Scenarios," Transactions on Emerging Telecommunications Technologies (ETT), vol. 33, no. 5, e4452, 2022, ISSN: 2161-3915. DOI: https://doi.org/10.1002/ett.4452.
14. F. Kuntke, M. Sinn, S. Linsner, and C. Reuter, "Low Power Wide Area Networks (LPWAN) für krisentaugliche Datenübertragung in landwirtschaftlichen Betrieben," in 41. GIL-Jahrestagung: Informatik in der Land-, Forst- und Ernährungswirtschaft, A. Meyer-Aurich, M. Gandorfer, C. Hoffmann, C. Weltzien, S. D. Bellingrath-Kimura, and H. Floto, Eds., Meyer-Aurich, Andreas Gandorfer, Markus Hoffmann, Christa Weltzien, Cornelia Bellingrath-Kimura, Sonoko D. Floto, Helga: Gesellschaft für Informatik, 2021, pp. 193–198. [Online]. Available: http://www.peasec.de/paper/2021/2021_KuntkeSinnLinsnerReuter_LowPowerWideAreaNetworks_GIL.pdf.
15. A. Bernardi, S. Becker, C. Struve, S. Linsner, C. Reuter, and G. Müller, "Erfolgsorientierte Dienstleistung: Neue Perspektiven für die landwirtschaftliche Arbeitsteilung durch hybride Dienstleistungen," in 41. GIL-Jahrestagung: Informatik in der Land-, Forst- und Ernährungswirtschaft, A. Meyer-Aurich, M. Gandorfer, C. Hoffmann, C. Weltzien, S. D. Bellingrath-Kimura, and H. Floto, Eds., Gesellschaft für Informatik, 2021, p. 37. [Online]. Available: https://peasec.de/paper/2021/2021_Bernardietal_ErfolgsorientierteDienstleistung_GIL.pdf.

16. S. Linsner, R. Varma, and C. Reuter, "Vulnerability Assessment in the Smart Farming Infrastructure through Cyberattacks," in 39. GIL-Jahrestagung: Informatik in der Land-, Forst- und Ernährungswirtschaft Fokus; Digitalisierung für landwirtschaftliche Betriebe in kleinstrukturierten Regionen – ein Widerspruch in sich?, Lecture Notes in Informatics (LNI), A. Meyer-Aurich, Ed., Wien, Austria: Gesellschaft für Informatik, 2019, pp. 119–124. [Online]. Available: http://gil-net.de/Publikationen/139_119.pdf.
17. A. Bernardi, C. Reuter, W. Schneider, S. Linsner, and M.-A. Kaufhold, "Hybride Dienstleistungen in digitalisierten Kooperationen in der Landwirtschaft," in 39. GIL-Jahrestagung: Informatik in der Land-, Forst- und Ernährungswirtschaft Fokus; Digitalisierung für landwirtschaftliche Betriebe in kleinstrukturierten Regionen – ein Widerspruch in sich?, Lecture Notes in Informatics (LNI), A. Meyer-Aurich, Ed., Vienna, Austria: Gesellschaft für Informatik, 2019, pp. 25–30. [Online]. Available: http://gil-net.de/Publikationen/139_25-30.pdf.
18. Statistisches Bundesamt, "Land- und Forstwirtschaft," in Statistisches Jahrbuch 2019, Wiesbaden: Statistisches Bundesamt (Destatis), 2019, ch. 19, pp. 487–520. [Online]. Available: https://www.destatis.de/DE/Themen/Querschnitt/Jahrbuch/statistisches-jahrbuch-2019-dl.pdf.
19. S. Alboaie and D. Cosovan, "Private data system enabling self-sovereign storage managed by executable choreographies," in Distributed Applications and Interoperable Systems: 17th IFIP WG 6.1 International Conference, DAIS 2017, Held as Part of the 12th International Federated Conference on Distributed Computing Techniques, DisCoTec 2017, Neuchâtel, Switzerland, June 19–22, 2017, Proceedings 17, Springer, 2017, pp. 83–98.
20. A. Fraser, "Land grab/data grab: Precision agriculture and its new horizons," The Journal of Peasant Studies, vol. 46, no. 5, pp. 893–912, 2019.
21. P. Grimm, "Social desirability bias," Wiley international encyclopedia of marketing, 2010.
22. S. Dechand, A. Naiakshina, A. Danilova, and M. Smith, "In encryption we dont trust: The effect of end-to-end encryption to the masses on user perception," in 2019 IEEE European Symposium on Security and Privacy (EuroS&P), 2019, pp. 401–415. DOI: https://doi.org/10.1109/EuroSP.2019.00037.
23. A. Hovav and F. F. Putri, "This is my device! why should i follow your rules? employees compliance with byod security policy," Pervasive and Mobile Computing, vol. 32, pp. 35–49, 2016, Mobile Security, Privacy and Forensics, ISSN: 1574-1192. DOI: https://doi.org/10.1016/j.pmcj.2016.06.007. [Online]. Available: https://www.sciencedirect.com/science/article/pii/S1574119216300724.
24. A. Calvó-Armengol and M. O. Jackson, "Peer pressure," Journal of the European Economic Association, vol. 8, no. 1, pp. 62–89, 2010.
25. A. Tandon, A. Dhir, S. Talwar, P. Kaur, and M. Mäntymäki, "Social media induced fear of missing out (fomo) and phubbing: Behavioural, relational and psychological outcomes," Technological Forecasting and Social Change, vol. 174, p. 121 149, 2022.
26. R. Von Solms and J. Van Niekerk, "From information security to cyber security," computers & security, vol. 38, pp. 97–102, 2013.
27. L. Barkhuus, "The mismeasurement of privacy: Using contextual integrity to reconsider privacy in hci," in Proceedings of the SIGCHI Conference on Human Factors in Computing Systems, 2012, pp. 367–376.

28. P. Hummel, M. Braun, M. Tretter, and P. Dabrock, "Data sovereignty: A review," Big Data & Society, vol. 8, no. 1, p. 2 053 951 720 982 012, 2021.
29. K. Niebert and H. Gropengießer, "Leitfadengestzte interviews," Methoden in der naturwissenschaftsdidaktischen Forschung, pp. 121–132, 2014.
30. J. Lazar, J. H. Feng, and H. Hochheiser, Research Methods in Human-Computer Interaction. Morgan Kaufmann, 2017, pp. 1–560, ISBN: 9780128053904.
31. D. Morgan, Focus Groups as Qualitative Research. 2455 Teller Road, Thousand Oaks California 91320 United States of America: SAGE Publications, Inc., 1997, vol. 16, pp. 32–46, ISBN: 9780761903437. DOI: https://doi.org/10.4135/9781412984287. [Online]. Available: http://methods.sagepub.com/book/focus-groups-as-qualitative-research.
32. E. Duque, G. Fonseca, H. Vieira, G. Gontijo, and L. Ishitani, "A systematic literature review on user centered design and participatory design with older people," in Proceedings of the 18th Brazilian Symposium on Human Factors in Computing Systems, ser. IHC '19, Vitória, Espırito Santo, Brazil: Association for Computing Machinery, 2019, ISBN: 9781450369718. DOI: https://doi.org/10.1145/3357155.3358471. [Online]. Available: https://doi.org/10.1145/3357155.3358471.
33. P. A. Heslin and U.-C. Klehe, "Self-efficacy," Encyclopedia Of Industrial/Organizational Psychology, SG Rogelberg, ed, vol. 2, pp. 705–708, 2006.
34. L. Brandimarte, A. Acquisti, and G. Loewenstein, "Misplaced confidences: Privacy and the control paradox," Social Psychological and Personality Science, vol. 4, pp. 340–347, 3 2013, ISSN: 19485506. DOI: https://doi.org/10.1177/1948550612455931.
35. I. D. Boer, E. Herder, and M. V. Lieshoud, "The illusion of control in privacy trade-offs: Does familiarity play a role?" In Adjunct Proceedings of the 29th ACM Conference on User Modeling, Adaptation and Personalization, Association for Computing Machinery, Inc, 2021, pp. 338–344, ISBN: 9781450383677. DOI: https://doi.org/10.1145/3450614.3464471.
36. Yu Gu and Tiaobin Jing, "The IOT research in supply chain management of fresh agricultural products," in 2011 2nd International Conference on Artificial Intelligence, Management Science and Electronic Commerce (AIMSEC), IEEE, 2011, pp. 7382–7385, ISBN: 978-1-4577-0535-9. DOI: https://doi.org/10.1109/AIMSEC.2011.6011477. [Online]. Available: http://ieeexplore.ieee.org/document/6011477/.
37. S. Wolfert, L. Ge, C. Verdouw, and M.-J. Bogaardt, "Big Data in Smart Farming – A review," Agricultural Systems, vol. 153, pp. 69–80, 2017, ISSN: 0308521X. DOI: https://doi.org/10.1016/j.agsy.2017.01.023. [Online]. Available: https://linkinghub.elsevier.com/retrieve/pii/S0308521X16303754.
38. D. Jea, I. Yap, and M. Srivastava, "User access of public shared devices in pervasive computing environments," in 2007 Joint Workshop on High Confidence Medical Devices, Software, and Systems and Medical Device Plug-and-Play Interoperability (HCMDSS-MDPnP 2007), IEEE, 2007, pp. 57–63.
39. H. Choi, J. Park, and Y. Jung, "The role of privacy fatigue in online privacy behavior," Computers in Human Behavior, vol. 81, pp. 42–51, 2018.
40. L. Schneidewind, S. Hörold, C. Mayas, H. Krömker, S. Falke, and T. Pucklitsch, "How personas support requirements engineering," in 2012 First International Workshop on Usability and Accessibility Focused Requirements Engineering (UsARE), Zurich, Switzerland: IEEE, 2012, pp. 1–5.
41. M. J. Culnan, " " how did they get my name?": An exploratory investigation of consumer attitudes toward secondary information use," MIS quarterly, pp. 341–363, 1993.

42. C. Braz, A. Seffah, and D. M'Raihi, "Designing a trade-off between usability and security: A metrics based-model," in Human-Computer Interaction – INTERACT 2007, C. Baranauskas, P. Palanque, J. Abascal, and S. D. J. Barbosa, Eds., Berlin, Heidelberg: Springer Berlin Heidelberg, 2007, pp. 114–126, ISBN: 978-3-540-74800-7.
43. O. Kulyk, S. Neumann, J. Budurushi, and M. Volkamer, "Nothing comes for free: How much usability can you sacrifice for security?" IEEE Security & Privacy, vol. 15, no. 3, pp. 24–29, 2017. DOI: https://doi.org/10.1109/MSP.2017.70.
44. S. A. Alsuhibany, "A proposed approach for handling the tradeoff between security, usability, and cost," in 2019 International Conference on Computer and Information Sciences (ICCIS), 2019, pp. 1–6. DOI: https://doi.org/10.1109/ICCISci.2019.8716447.
45. T. Wu, R. Zhang, W. Ma, S. Wen, X. Xia, C. Paris, S. Nepal, and Y. Xiang, "What risk? i don't understand. an empirical study on users' understanding of the terms used in security texts," in Proceedings of the 15th ACM Asia Conference on Computer and Communications Security, New York, NY, USA: Association for Computing Machinery, 2020, pp. 248–262.
46. S. Barth, D. Ionita, M. D. De Jong, P. H. Hartel, and M. Junger, "Privacy rating: A user-centered approach for visualizing data handling practices of online services," IEEE transactions on professional communication, vol. 64, no. 4, pp. 354–373, 2021.
47. M. Gandorfer, S. Schleicher, S. Heuser, and J. Pfeiffer, "Landwirtschaft 4.0 – Digitalisierung und ihre Herausforderungen," pp. 9–19, 2017. [Online]. Available: https://www.lfl.bayern.de/mam/cms07/ilt/dateien/digitalisierung%5C_und%5C_ihre%5C_herausforderungen.pdf.
48. A.-T. Braun, E. Colangelo, and T. Steckel, "Farming in the Era of Industrie 4.0," Procedia CIRP, vol. 72, pp. 979–984, 2018, ISSN: 22128271. DOI: https://doi.org/10.1016/j.procir.2018.03.176. [Online]. Available: https://linkinghub.elsevier.com/retrieve/pii/S2212827118303342.
49. A. Tzounis, N. Katsoulas, T. Bartzanas, and C. Kittas, "Internet of Things in agriculture, recent advances and future challenges," Biosystems Engineering, vol. 164, pp. 31–48, 2017, ISSN: 15375110. DOI: https://doi.org/10.1016/j.biosystemseng.2017.09.007. [Online]. Available: http://www.sciencedirect.com/science/article/pii/S1537511017302544.
50. R. Lokers, R. Knapen, S. Janssen, Y. van Randen, and J. Jansen, "Analysis of Big Data technologies for use in agro-environmental science," Environmental Modelling & Software, vol. 84, pp. 494–504, 2016, ISSN: 13648152. DOI: https://doi.org/10.1016/j.envsoft.2016.07.017. [Online]. Available: http://www.sciencedirect.com/science/article/pii/S1364815216304194.
51. W. Melicher, M. Sharif, J. Tan, L. Bauer, M. Christodorescu, and P. G. Leon, "(Do Not) Track Me Sometimes: Users' Contextual Preferences for Web Tracking," Proceedings on Privacy Enhancing Technologies, vol. 2016, no. 2, pp. 135–154, 2016. DOI: https://doi.org/10.1515/popets-2016-0009. [Online]. Available: https://content.sciendo.com/view/journals/popets/2016/2/article-p135.xml.
52. J. L. Ferris, "Data privacy and protection in the agriculture industry: Is federal regulation necessary," Minn. JL Sci. & Tech., vol. 18, p. 309, 2017. [Online]. Available: https://scholarship.law.umn.edu/cgi/viewcontent.cgi?article=1422&context=mjlst.
53. L. Aldehoff, M. Dankenbring, and C. Reuter, "Renouncing privacy in crisis management? people's view on social media monitoring and surveillance," Proceedings of the

Information Systems for Crisis Response and Management (ISCRAM), pp. 1184–1197, 2019.
54. R. Kamath, "Food Traceability on Blockchain: Walmart's Pork and Mango Pilots with IBM," The Journal of the British Blockchain Association, vol. 1, no. 1, pp. 1–12, 2018, ISSN: 25163949. DOI: https://doi.org/10.31585/jbba-1-1-(10)2018. [Online]. Available: https://jbba.scholasticahq.com/article/3712-food-traceability-on-blockchain-walmart-s-pork-and-mango-pilots-with-ibm.
55. O. Bermeo-Almeida, M. Cardenas-Rodriguez, T. Samaniego-Cobo, E. Ferruzola-Gómez, R. Cabezas-Cabezas, and W. Bazán-Vera, "Blockchain in agriculture: A systematic literature review," in Technologies and Innovation, R. Valencia-García, G. Alcaraz-Mármol, J. Del Cioppo-Morstadt, N. Vera-Lucio, and M. Bucaram-Leverone, Eds., Cham: Springer International Publishing, 2018, pp. 44–56, ISBN: 978-3-030-00940-3.
56. L. Ge, C. Brewster, J. Spek, A. Smeenk, J. Top, F. van Diepen, B. Klaase, C. Graumans, and M. de Ruyter de Wildt, "Blockchain for agriculture and food," Tech. Rep., 2017. DOI: https://doi.org/10.18174/426747. [Online]. Available: http://library.wur.nl/WebQuery/wurpubs/530264.
57. O. Elijah, T. A. Rahman, I. Orikumhi, C. Y. Leow, and M. N. Hindia, "An overview of internet of things (iot) and data analytics in agriculture: Benefits and challenges," IEEE Internet of Things Journal, vol. 5, no. 5, pp. 3758–3773, 2018.
58. K. Rosskopf and P. Wagner, "Vom Daten- zum Wissensmanagement : Wofür verwenden Landwirte einen Computer ?" In GIL Jahrestagung, 2006, pp. 225–228.
59. M. Pinaki and V. K. Tewari, "Present status of precision farming: A review.," International Journal of Agricultural Research, vol. 5, no. 12, pp. 1124–1133, 2010. DOI: https://doi.org/10.3923/ijar.2007.1.10.
60. R. Finger, S. M. Swinton, N. El Benni, and A. Walter, "Precision Farming at the Nexus of Agricultural Production and the Environment," Annual Review of Resource Economics, vol. 11, no. 1, pp. 313–335, 2019. DOI: https://doi.org/10.1146/annurev-resource-100518-093929. [Online]. Available: https://doi.org/10.1146/annurev-resource-100518-093929.
61. N. Pelletier, E. Audsley, S. Brodt, T. Garnett, P. Henriksson, A. Kendall, K. J. Kramer, D. Murphy, T. Nemecek, and M. Troell, "Energy Intensity of Agriculture and Food Systems," Annual Review of Environment and Resources, vol. 36, no. 1, pp. 223–246, 2011. DOI: https://doi.org/10.1146/annurev-environ-081710-161014. [Online]. Available: https://doi.org/10.1146/annurev-environ-081710-161014.
62. M. Shepherd, J. A. Turner, B. Small, and D. Wheeler, "Priorities for science to overcome hurdles thwarting the full promise of the 'digital agriculture'revolution," Journal of the Science of Food and Agriculture, 2018.
63. P. E. Naeini, S. Bhagavatula, H. Habib, M. Degeling, L. Bauer, L. F. Cranor, and N. Sadeh, "Privacy expectations and preferences in an iot world," in Thirteenth Symposium on Usable Privacy and Security (SOUPS 2017), Santa Clara, CA: USENIX Association, 2017, pp. 399–412, ISBN: 978-1-931971-39-3. [Online]. Available: https://www.usenix.org/conference/soups2017/technical-sessions/presentation/naeini.
64. M. E. Sykuta, "Big Data in Agriculture: Property Rights, Privacy and Competition in Ag Data Services," International Food and Agribusiness Management Review, Volume 19, no. 1030-2016-83141, p. 18, 2016. DOI: https://doi.org/10.22004/ag.econ.240696. [Online]. Available: http://ageconsearch.umn.edu/record/240696.

65. M. Nery, R. Santos, W. Santos, V. Lourenco, and M. Moreno, "Facing digital agriculture challenges with knowledge engineering," in 2018 First International Conference on Artificial Intelligence for Industries (AI4I), IEEE, 2018, pp. 118–119.
66. A. Fleming, E. Jakku, L. Lim-Camacho, B. Taylor, and P. Thorburn, "Is big data for big farming or for everyone? perceptions in the Australian grains industry," Agronomy for sustainable development, vol. 38, no. 3, p. 24, 2018.
67. A. Zhang, E. Jakku, R. Llewellyn, and E. Bake, "Surveying the needs and drivers for digital agriculture in Australia," Farm Policy Journal, vol. 15, no. 1, pp. 25–39, 2018.
68. S. Fountas, S. Blackmore, D. Ess, S. Hawkins, G. Blumhoff, J. Lowenberg-Deboer, and C. Sorensen, "Farmer experience with precision agriculture in denmark and the us eastern corn belt," Precision Agriculture, vol. 6, no. 2, pp. 121–141, 2005.
69. I. Carbonell, "The ethics of big data in big agriculture," Internet Policy Review, vol. 5, no. 1, 2016.
70. M. Kernecker, A. Knierim, A. Wurbs, T. Kraus, and F. Borges, "Experience versus expectation: farmers' perceptions of smart farming technologies for cropping systems across Europe," Precision Agriculture, vol. 21, no. 1, pp. 34–50, 2020, ISSN: 1385-2256. DOI: https://doi.org/10.1007/s11119-019-09651-z [Online]. Available: http://link.springer.com/10.1007/s11119-019-09651-z
71. Á. Regan, S. Green, P. Maher, et al., "Smart farming in ireland: Anticipating positive and negative impacts through a qualitative study of risk and benefit perceptions amongst expert actors in the irish agri-food sector," in Proceedings of the 13th European International Farm Systems Association Symposium, Chania, Greece, 2018, pp. 1–5. [Online]. Available: http://ifsa.boku.ac.at/cms/fileadmin/Proceeding2018/Theme4_Regan.pdf.
72. S. Cravotta and M. Grottke, "Digitalization in German family firms–some preliminary insights," Journal of Evolutionary Studies in Business, vol. 4, no. 1, pp. 1–25, 2019. DOI: https://doi.org/10.1344/jesb2019.1.j051.
73. T. Biselli and C. Reuter, "On the relationship between it privacy and security behavior: A survey among German private users," Wirtschaftsinformatik 2021 Proceedings, pp. 1–17, 2021.
74. N. Malkin, J. Deatrick, A. Tong, P. Wijesekera, S. Egelman, and D. Wagner, "Privacy attitudes of smart speaker users," Proceedings on Privacy Enhancing Technologies, vol. 2019, no. 4, pp. 250–271, 1Oct. 2019. DOI: https://doi.org/10.2478/popets-2019-0068. [Online]. Available: https://content.sciendo.com/view/journals/popets/2019/4/article-p250.xml.
75. V. Zimmermann, P. Gerber, K. Marky, L. Böck, and F. Kirchbuchner, "Assessing users privacy and security concerns of smart home technologies," i-com, vol. 18, no. 3, pp. 197–216, 2019.
76. C. Han, I. Reyes, Á. Feal, J. Reardon, P. Wijesekera, N. Vallina-Rodriguez, A. Elazari, K. A. Bamberger, and S. Egelman, "The price is (not) right: Comparing privacy in free and paid apps," Proceedings on Privacy Enhancing Technologies, vol. 2020, no. 3, pp. 222–242, 1Jul. 2020. DOI: https://doi.org/10.2478/popets-2020-0050. [Online]. Available: https://content.sciendo.com/view/journals/popets/2020/3/article-p222.xml.
77. D. Smullen, Y. Feng, S. A. Zhang, and N. Sadeh, "The best of both worlds: Mitigating trade-offs between accuracy and user burden in capturing mobile app privacy preferences," Proceedings on Privacy Enhancing Technologies, vol. 2020, no. 1, pp. 195–215, 1Jan. 2020. DOI: https://doi.org/10.2478/popets-2020-0011. [Online]. Available: https://content.sciendo.com/view/journals/popets/2020/1/article-p195.xml.

78. K. P. Coopamootoo and T. Groß, "Why privacy is all but forgotten," Proceedings on Privacy Enhancing Technologies, vol. 2017, no. 4, pp. 97–118, 1Oct. 2017. DOI: https://doi.org/10.1515/popets-2017-0040. [Online]. Available: https://content.sciendo.com/view/journals/popets/2017/4/article-p97.xml.
79. S. S. Hassan, C. Reuter, and L. Bzhalava, "Perception or capability? – an empirical investigation of the factors influencing the adoption of social media and public cloud in German SMEs," International Journal of Innovation Management, pp. 1–29, 2020. DOI: https://doi.org/10.1142/S136391962150002X.
80. Eurostat, Agriculture: Labour force by sex, legal statusof holding and working time, 2016. [Online]. Available: https://ec.europa.eu/eurostat/web/agriculture/data/database.
81. J. M. Corbin and A. Strauss, "Grounded theory research: Procedures, canons, and evaluative criteria," Qualitative sociology, vol. 13, no. 1, pp. 3–21, 1990.
82. H. Doll, F. Fasterding, and K. Klare, "Auswirkungen des landwirtschaftlichen Erbrechts auf den agrarstrukturellen Wandel in Deutschland," German Journal of Agricultural Economics, no. 670-2016-45544, p. 5, 2001. DOI: https://doi.org/10.22004/ag.econ.98863. [Online]. Available: http://ageconsearch.umn.edu/record/98863.
83. C. Reuter, Information Technology for Peace and Security - IT-Applications and Infrastructures in Conflicts, Crises, War, and Peace, C. Reuter, Ed. Wiesbaden: Springer Fachmedien Wiesbaden, 2019, pp. 1–424, ISBN: 978-3-658-25651-7. DOI: https://doi.org/10.1007/978-3-658-25652-4. [Online]. Available: http://link.springer.com/10.1007/978-3-658-25652-4.
84. H. Saleem and M. Naveed, "SoK: Anatomy of Data Breaches," Proceedings on Privacy Enhancing Technologies, vol. 2020, no. 4, pp. 153–174, 2020, ISSN: 2299-0984. DOI: https://doi.org/10.2478/popets-2020-0067. [Online]. Available: https://content.sciendo.com/view/journals/popets/2020/4/article-p153.xml.
85. D. Froelicher, P. Egger, J. S. Sousa, J. L. Raisaro, Z. Huang, C. Mouchet, B. Ford, and J.-P. Hubaux, "UnLynx: A Decentralized System for Privacy-Conscious Data Sharing," Proceedings on Privacy Enhancing Technologies, vol. 2017, no. 4, pp. 232–250, 2017, ISSN: 2299-0984. DOI: https://doi.org/10.1515/popets-2017-0047. [Online]. Available: https://content.sciendo.com/view/journals/popets/2017/4/article-p232.xml.
86. A. Pham, I. Dacosta, B. Jacot-Guillarmod, K. Huguenin, T. Hajar, F. Tramèr, V. Gligor, and J.-P. Hubaux, "PrivateRide: A Privacy-Enhanced Ride-Hailing Service," Proceedings on Privacy Enhancing Technologies, vol. 2017, no. 2, pp. 38–56, 2017. DOI: https://doi.org/10.1515/popets-2017-0015. [Online]. Available: https://content.sciendo.com/view/journals/popets/2017/2/article-p38.xml.
87. D. B. (DBV), "Positionspapier des präsidiums des deutschen bauernverbandes vom 13. september 2016 zu landwirtschaft 4.0 – chancen und handlungsbedarf," 2016. [Online]. Available: https://media.repro-mayr.de/34/661134.pdf.
88. A. Sanders, C. Elangeswaran, and J. P. Wulfsberg, "Industry 4.0 implies lean manufacturing: Research activities in industry 4.0 function as enablers for lean manufacturing," Journal of Industrial Engineering and Management (JIEM), vol. 9, no. 3, pp. 811–833, 2016.
89. C. Reuter, Information Technology for Peace and Security – IT Applications and Infrastructures in Conflicts, Crises, War, and Peace, C. Reuter, Ed. Springer Vieweg, 2019, ISBN: 9783658256517.
90. C. Reuter, Sicherheitskritische Mensch-Computer-Interaktion – Interaktive Technologien und Soziale Medien im Krisen- und Sicherheitsmanagement, C. Reuter, Ed. Springer Vieweg, 2018, ISBN: 978-3-658-19522-9.

91. C. Reuter, W. Schneider, D. Eberz, M. Bayer, D. Hartung, and C. Kaygusuz, "Resiliente Digitalisierung der kritischen Infrastruktur Landwirtschaft - mobil, dezentral, ausfallsicher," in Mensch und Computer 2018 - Workshopband, R. Dachselt and G. Weber, Eds., Bonn: Gesellschaft für Informatik e.V., 2018.
92. S. A. Abeyratne and R. P. Monfared, "Blockchain ready manufacturing supply chain using distributed ledger," 2016.
93. B. A. Aubert, A. Schroeder, and J. Grimaudo, "IT as enabler of sustainable farming: An empirical analysis of farmers' adoption decision of precision agriculture technology," Decision Support Systems, vol. 54, no. 1, pp. 510–520, 2012, ISSN: 01679236. DOI: https://doi.org/10.1016/j.dss.2012.07.002. [Online]. Available: http://dx.doi.org/10.1016/j.dss.2012.07.002.
94. M. Paustian and L. Theuvsen, "Adoption of precision agriculture technologies by german crop farmers," Precision agriculture, vol. 18, no. 5, pp. 701–716, 2017.
95. C.-F. V. Hobe, M. Michels, W. Fecke, O. Mußhoff, P. Johann, and W. V. Ahlefeld, "Wie kommunizieren Landwirte in Zeiten der Digitalisierung? Ergebnisse und Diskussion," pp. 269–274, 2019.
96. V. Brühl, "Bitcoins, Blockchain und Distributed Ledgers," Wirtschaftsdienst, vol. 97, no. 2, pp. 135–142, 2017, ISSN: 0043-6275. DOI: https://doi.org/10.1007/s10273-017-2096-3. [Online]. Available: http://link.springer.com/10.1007/s10273-017-2096-3.
97. C. Maple and J. Jackson, "Selecting effective blockchain solutions," in European Conference on Parallel Processing, Springer, 2018, pp. 392–403.
98. T. Koens and E. Poll, "The drivers behind blockchain adoption: The rationality of irrational choices," Lecture Notes in Computer Science, vol. 11339, pp. 535–546, 2018, ISSN: 16113349. DOI: https://doi.org/10.1007/978-3-030-10549-5_42.
99. K. Wüst and A. Gervais, "Do you need a blockchain?" 2018 Crypto Valley Conference on Blockchain Technology (CVCBT), pp. 45–54, 2017.
100. J. Kitzinger, "Qualitative [Research: Introducing focus groups," BMJ, no. 7000, p. 299, ISSN: 14685833. DOI:https://doi.org/10.1136/bmj.311.7000.299.
101. H. Kawamura and D. L. Morgan, "Focus Groups as Qualitative Research," The Modern Language Journal, vol. 82, no. 4, p. 594, 2006, ISSN: 00267902. DOI: https://doi.org/10.2307/330238.
102. B. für Ernährung und Landwirtschaft, "Landwirtschaft in der eu: Altersgliederung der landwirtschaftlichen betriebsinhaber," Tabellen zur Landwirtschaft, 2013. [Online]. Available: https://www.bmel-statistik.de/landwirtschaft/tabellen-zur-landwirtschaft/#c7002.
103. J.-P. Jussila, "Reconciling the conflict between the immutability of public and permissionless blockchain technology and the right to erasure under Article 17 of the General Data Protection Regulation," en, 2018.
104. C. Paunov and S. Planes-Satorra, "How are digital technologies changing innovation?" OECD Science, Technology and Industry Policy Papers, no. 74, 2019. DOI: https://doi.org/10.1787/67bbcafe-en. [Online]. Available: https://www.oecd-ilibrary.org/content/paper/67bbcafe-en.
105. E. Steinbrink, L. Reichert, M. Mende, and C. Reuter, "Digital Privacy Perceptions of Asylum Seekers in Germany - An Empirical Study about Smartphone Usage during the Flight," Proceedings of the ACM: Human Computer Interaction (PACM): Computer-Supported Cooperative Work and Social Computing, vol. 5, no. CSCW2, 2021. DOI: https://doi.org/10.1145/3479526. [Online].

Available: https://www.peasec.de/paper/2021/2021_SteinbrinkReichertMendeReuter_DigitalPrivacyPerceptionAsylumSeekers_CSCW.pdf.
106. C. Atik and B. Martens, "Competition problems and governance of non-personal agricultural machine data: Comparing voluntary initiatives in the US and EU," Journal of Intellectual Property, Information Technology and E-Commerce Law: JIPITEC, vol. 12, no. 3, pp. 370–396, 2021.
107. M. Gupta, M. Abdelsalam, S. Khorsandroo, and S. Mittal, "Security and Privacy in Smart Farming: Challenges and Opportunities," IEEE Access, vol. 8, pp. 34 564–34 584, 2020, ISSN: 2169-3536. DOI: https://doi.org/10.1109/ACCESS.2020.2975142. [Online]. Available: https://ieeexplore.ieee.org/document/9003290/.
108. Eurostat, Agriculture: Labour force by sex, legal status of holding and working time, 2018. [Online]. Available: https://ec.europa.eu/eurostat/web/agriculture/data/database.
109. S. Rotz, E. Duncan, M. Small, J. Botschner, R. Dara, I. Mosby, M. Reed, and E. D. Fraser, "The Politics of Digital Agricultural Technologies: A Preliminary Review," Sociologia Ruralis, vol. 59, no. 2, pp. 203–229, 2019, ISSN: 0038-0199. DOI: https://doi.org/10.1111/soru.12233. [Online]. Available: https://onlinelibrary.wiley.com/doi/10.1111/soru.12233.
110. B. C. Runck, A. Joglekar, K. A. T. Silverstein, C. Chan-Kang, P. G. Pardey, and J. C. Wilgenbusch, "Digital agriculture platforms: Driving data-enabled agricultural innovation in a world fraught with privacy and security concerns," Agronomy Journal, vol. 113, no. 6, pp. 5412–5423, 2021, ISSN: 0002-1962. DOI: https://doi.org/10.1002/agj2.20873. [Online]. Available: https://onlinelibrary.wiley.com/doi/10.1002/agj2.20791, https://onlinelibrary.wiley.com/doi/10.1002/agj2.20873.
111. M. C. Annosi, F. Brunetta, A. Monti, and F. Nati, "Is the trend your friend? An analysis of technology 4.0 investment decisions in agricultural SMEs," Computers in Industry, vol. 109, pp. 59–71, 2019, ISSN: 01663615. DOI: https://doi.org/10.1016/j.compind.2019.04.003. [Online]. Available: https://www.sciencedirect.com/science/article/pii/S0166361518305992 https://linkinghub.elsevier.com/retrieve/pii/S0166361518305992
112. J. Becker, M. Heddier, A. Öksuz, and R. Knackstedt, "The effect of providing visualizations in privacy policies on trust in data privacy and security," in 2014 47th Hawaii International Conference on System Sciences, 2014, pp. 3224–3233. DOI: https://doi.org/10.1109/HICSS.2014.399.
113. K. Hartwig and C. Reuter, "Nudging Users Towards Better Security Decisions in Password Creation Using Whitebox-based Multidimensional Visualizations," Behaviour & Information Technology (BIT), 2021. DOI: https://doi.org/10.1080/0144929X.2021.1876167. [Online]. Available: https://peasec.de/paper/2021/2021_HartwigReuter_WhiteboxMultidimensionalNudges_BIT.pdf.
114. K. Hartwig and C. Reuter, "Nudge or Restraint: How do People Assess Nudging in Cybersecurity - A Representative Study in Germany," in European Symposium on Usable Security (EuroUSEC), Karlsruhe, Germany: ACM, 2021, pp. 141–150. DOI: https://doi.org/10.1145/3481357.3481514. [Online]. Available: https://peasec.de/paper/2021/2021_HartwigReuter_NudgingCybersecurityRepresentativeStudy_EuroUSEC.pdf.
115. N. Gerber, P. Gerber, and M. Volkamer, "Explaining the privacy paradox: A systematic review of literature investigating privacy attitude and behavior," Computers & Security, vol. 77, pp. 226–261, 2018. [Online]. Available: https://madoc.bib.uni-mannheim.de/50606/.

116. L. L. Iacono, M. Smith, and P. L. Gorski, "Werkzeuge für usable (cyber-)security," in Sicherheitskritische Mensch-Computer-Interaktion: Interaktive Technologien und Soziale Medien im Krisen- und Sicherheitsmanagement, C. Reuter, Ed. Wiesbaden: Springer Fachmedien Wiesbaden, 2021, pp. 111–130, ISBN: 978-3-658-32795-8. DOI: https://doi.org/10.1007/978-3-658-32795-8_6. [Online]. Available: https://doi.org/10.1007/978-3-658-32795-8_6.
117. R. Iannella and A. Finden, "Privacy awareness: Icons and expression for social networks," in Proceedings of the 8th Virtual Goods Workshop and the 6th ODRL Workshop, 2010, pp. 1–15.
118. L.-E. Holtz, K. Nocun, and M. Hansen, "Towards Displaying Privacy Information with Icons," in Privacy and Identity Management for Life, S. Fischer-Hübner, P. Duquenoy, M. Hansen, R. Leenes, and G. Zhang, Eds., Berlin, Heidelberg: Springer Berlin Heidelberg, 2011, pp. 338–348, ISBN: 978-3-642-20769-3.
119. K. Ghazinour, M. Majedi, and K. Barker, "A model for privacy policy visualization," in 2009 33rd Annual IEEE International Computer Software and Applications Conference, vol. 2, 2009, pp. 335–340. DOI:https://doi.org/10.1109/COMPSAC.2009.156.
120. Y. Wang, L. Gou, A. Xu, M. X. Zhou, H. Yang, and H. Badenes, "Veilme: An interactive visualization tool for privacy configuration of using personality traits," in Proceedings of the 33rd Annual ACM Conference on Human Factors in Computing Systems, ser. CHI '15, Seoul, Republic of Korea: Association for Computing Machinery, 2015, pp. 817–826, ISBN: 9781450331456. DOI: https://doi.org/10.1145/2702123.2702293. [Online]. Available: https://doi.org/10.1145/2702123.2702293.
121. H. R. Lipford, A. Besmer, and J. Watson, "Understanding privacy settings in facebook with an audience view," in Proceedings of the 1st Conference on Usability, Psychology, and Security, ser. UPSEC'08, San Francisco, California: USENIX Association, 2008.
122. W. Tolone, G.-J. Ahn, T. Pai, and S.-P. Hong, "Access control in collaborative systems," ACM Comput. Surv., vol. 37, no. 1, pp. 29–41, 2005, ISSN: 0360-0300. DOI: https://doi.org/10.1145/1057977.1057979. [Online]. Available: https://doi.org/10.1145/1057977.1057979.
123. F. Paci, A. Squicciarini, and N. Zannone, "Survey on access control for community-centered collaborative systems," ACM Comput. Surv., vol. 51, no. 1, 2018, ISSN: 0360-0300. DOI: https://doi.org/10.1145/3146025. [Online]. Available: https://doi.org/10.1145/3146025.
124. J. M. Kizza, "Access Control and Authorization," in Computer Network Security, New York: Springer-Verlag, 2015, ch. 9, pp. 185–204. DOI: https://doi.org/10.1007/978-1-4471-6654-2_9. [Online]. Available: http://link.springer.com/10.1007/0-387-25228-2_8 http://link.springer.com/10.1007/978-1-4471-6654-2_9.
125. B. W. Lampson, "Protection," SIGOPS Oper. Syst. Rev., vol. 8, no. 1, pp. 18–24, 1974, ISSN: 0163-5980. DOI:https://doi.org/10.1145/775265.775268. [Online]. Available: https://doi.org/10.1145/775265.775268.
126. H. Hu, G.-J. Ahn, and J. Jorgensen, "Detecting and resolving privacy conflicts for collaborative data sharing in online social networks," in Proceedings of the 27th Annual Computer Security Applications Conference, ser. ACSAC '11, Orlando, Florida, USA: Association for Computing Machinery, 2011, pp. 103–112, ISBN: 9781450306720. DOI: https://doi.org/10.1145/2076732.2076747. [Online]. Available: https://doi.org/10.1145/2076732.2076747.

127. R. W. Reeder, L. Bauer, L. F. Cranor, M. K. Reiter, K. Bacon, K. How, and H. Strong, "Expandable grids for visualizing and authoring computer security policies," ser. CHI '08, Florence, Italy: Association for Computing Machinery, 2008, pp. 1473–1482, ISBN: 9781605580111. DOI: https://doi.org/10.1145/1357054.1357285. [Online]. Available: https://doi.org/10.1145/1357054.1357285.
128. M. Kolomeets, A. Chechulin, I. Kotenko, and I. Saenko, "Access control visualization using triangular matrices," in 2019 27th Euromicro International Conference on Parallel, Distributed and Network-Based Processing (PDP), 2019, pp. 348–355. DOI: https://doi.org/10.1109/EMPDP.2019.8671578.
129. J. Angulo, S. Fischer-Hübner, T. Pulls, and E. Wästlund, "Usable Transparency with the Data Track," in Proceedings of the 33rd Annual ACM Conference Extended Abstracts on Human Factors in Computing Systems, New York, NY, USA: ACM, 2015, pp. 1803–1808, ISBN: 9781450331463. DOI: https://doi.org/10.1145/2702613.2732701. [Online]. Available: https://dl.acm.org/doi/10.1145/2702613.2732701.
130. A. Disney, trust-hub: using graph technologies to power personal data compliance, Accessed on 2022-01-18, 2018. [Online]. Available: https://cambridge-intelligence.com/trust-hub-using-graph-technologies-to-power-personal-data-compliance/.
131. C. Bier, K. Kühne, and J. Beyerer, "Privacyinsight: The next generation privacy dashboard," in Annual Privacy Forum, Springer, 2016, pp. 135–152.
132. M. Wang, "Accessible Access Control: A Visualization Ssystem for Access Control Policy Management," PhD dissertation, Michigan Technological University, Houghton, Michigan, 2019. DOI: https://doi.org/10.37099/mtu.dc.etdr/841. [Online]. Available: https://digitalcommons.mtu.edu/etdr/841.
133. J.-K. Chou, C. Bryan, and K.-L. Ma, "Privacy preserving visualization for social network data with ontology information," in 2017 IEEE Pacific Visualization Symposium (PacificVis), 2017, pp. 11–20. DOI: https://doi.org/10.1109/PACIFICVIS.2017.8031573.
134. D. Wilkinson, P. Bahirat, M. Namara, J. Lyu, A. Alsubhi, J. Qiu, P. Wisniewski, and B. P. Knijnenburg, "Privacy at a Glance: The User-Centric Design of Glanceable Data Exposure Visualizations," Proceedings on Privacy Enhancing Technologies, vol. 2020, no. 2, pp. 416–435, 2020, ISSN: 2299-0984. DOI: https://doi.org/10.2478/popets-2020-0034. [Online]. Available: https://www.sciendo.com/article/10.2478/popets-2020-0034.
135. A. Bagnato, P. Silva, A. S. Alaqra, and O. Ermis, "Workshop on Privacy Challenges in Public and Private Organizations," in Privacy and Identity Management. Data for Better Living: AI and Privacy: 14th IFIP WG 9.2, 9.6/11.7, 11.6/SIG 9.2.2 International Summer School, Windisch, Switzerland, August 19–23, 2019, Revised Selected Papers, M. Friedewald, M. Önen, E. Lievens, S. Krenn, and S. Fricker, Eds. Cham: Springer International Publishing, 2020, pp. 82–89, ISBN: 978-3-030-42504-3. DOI: https://doi.org/10.1007/978-3-030-42504-3_6. [Online]. Available: https://doi.org/10.1007/978-3-030-42504-3_6.
136. J. Rode, C. Johansson, P. DiGioia, R. S. Filho, K. Nies, D. H. Nguyen, J. Ren, P. Dourish, and D. Redmiles, "Seeing further: Extending Visualization as a Basis for Usable Security," in Proceedings of the second symposium on Usable privacy and security - SOUPS '06, New York, New York, USA: ACM Press, 2006, p. 145, ISBN: 1595934480. DOI:https://doi.org/10.1145/1143120.1143138. [Online]. Available: http://portal.acm.org/citation.cfm?doid=1143120.1143138.
137. D. Feth and H. Schmitt, "Requirement and quality models for privacy dashboards," in 2020 IEEE 7th International Workshop on Evolving Security Privacy Requirements

Engineering (ESPRE), 2020, pp. 1–6. DOI: https://doi.org/10.1109/ESPRE51200.2020.00006.

138. J. Tolsdorf, M. Fischer, and L. Lo Iacono, "A Case Study on the Implementation of the Right of Access in Privacy Dashboards," in Privacy Technologies and Policy, N. Gruschka, L. F. C. Antunes, K. Rannenberg, and P. Drogkaris, Eds., Cham: Springer International Publishing, 2021, pp. 23–46, ISBN: 978-3-030-76663-4. DOI: https://doi.org/10.1007/978-3-030-76663-4_2.
139. K. Vaniea, Q. Ni, L. Cranor, and E. Bertino, "Access control policy analysis and visualization tools for security professionals," in SOUPS Workshop (USM), 2008, pp. 7–15.
140. J. Kolter, M. Netter, and G. Pernul, "Visualizing past personal data disclosures," in 2010 International Conference on Availability, Reliability and Security, 2010, pp. 131–139. DOI:https://doi.org/10.1109/ARES.2010.51.
141. S. van der Burg, L. Wiseman, and J. Krkeljas, "Trust in farm data sharing: reflections on the EU code of conduct for agricultural data sharing," Ethics and Information Technology, vol. 23, no. 3, pp. 185–198, 2021, ISSN: 1572-8439. DOI: https://doi.org/10.1007/s10676-020-09543-1. [Online]. Available: https://doi.org/10.1007/s10676-020-09543-1.
142. C. Parker and M. Sinclair, "User-centred design does make a difference. The case of decision support systems in crop production," Behaviour & Information Technology, vol. 20, no. 6, pp. 449–460, 2001, ISSN: 0144-929X. DOI: https://doi.org/10.1080/01449290110089570. [Online]. Available: http://www.tandfonline.com/doi/abs/10.1080/01449290110089570.
143. R. W. Reeder, L. Bauer, L. F. Cranor, M. K. Reiter, and K. Vaniea, "More than skin deep," in Proceedings of the SIGCHI Conference on Human Factors in Computing Systems, New York, NY, USA: ACM, 2011, pp. 2065–2074, ISBN: 9781450302289. DOI: https://doi.org/10.1145/1978942.1979243. [Online]. Available: https://dl.acm.org/doi/10.1145/1978942.1979243.
144. S. Pedersen, R. Ferguson, and R. Lark, "A multinational survej of precision farming early adopters," English, International Journal of Agricultural Management, vol. 11, no. 3, pp. 147–162, 2001, ISSN: 0014-8059.
145. J. Muñoz-Arteaga, R. M. González, M. V. Martin, J. Vanderdonckt, F. Álvarez-Rodriguez, and J. G. Calleros, "A method to design information security feedback using patterns and hci-security criteria," in Computer-Aided Design of User Interfaces VI, V. Lopez Jaquero, F. Montero Simarro, J. P. Molina Masso, and J. Vanderdonckt, Eds., London: Springer London, 2009, pp. 283–294, ISBN: 978-1-84882-206-1.
146. S. Garfinkel, "Design principles and patterns for computer systems that are simultaneously secure and usable," Ph.D. dissertation, Massachusetts Institute of Technology, 2005.
147. J. Brooke et al., "Sus-a quick and dirty usability scale," Usability evaluation in industry, vol. 189, no. 194, pp. 4–7, 1996.
148. M. Schrepp, A. Hinderks, and J. Thomaschewski, "Design and evaluation of a short version of the user experience questionnaire (ueq-s)," International Journal of Interactive Multimedia and Artificial Intelligence, 4 (6), 103-108., 2017.
149. T. Franke, C. Attig, and D. Wessel, "A personal resource for technology interaction:development and validation of the affinity for technology interaction (ati) scale,"

International Journal of Human-Computer Interaction, 2018. DOI: https://doi.org/10.1080/10447318.2018.1456150.
150. S. Holm, "A simple sequentially rejective multiple test procedure," Scandinavian journal of statistics, pp. 65–70, 1979.
151. J. Mariano, S. Marques, M. R. Ramos, F. Gerardo, C. L. da Cunha, A. Girenko, J. Alexandersson, B. Stree, M. Lamanna, M. Lorenzatto, L. P. Mikkelsen, U. Bundgård-Jørgensen, S. Rêgo, and H. de Vries, "Too old for technology? Stereotype threat and technology use by older adults," Behaviour & Information Technology, pp. 1–12, 2021, ISSN: 0144-929X. DOI: https://doi.org/10.1080/0144929X.2021.1882577. [Online]. Available: https://www.tandfonline.com/doi/full/10.1080/0144929X.2021.1882577.
152. T. Riebe, M.-A. Kaufhold, and C. Reuter, "The impact of organizational structure and technology use on collaborative practices in computer emergency response teams: An empirical study," Proceedings of the ACM on Human-Computer Interaction, vol. 5, no. CSCW2, pp. 1–30, 2021.
153. I. Goldberg, D. Wagner, and E. Brewer, "Privacy-enhancing technologies for the internet," in Proceedings IEEE COMPCON 97. Digest of Papers, San Jose, CA, USA: IEEE, 1997, pp. 103–109. DOI: https://doi.org/10.1109/CMPCON.1997.584680.
154. I. Goldberg, "Privacy-enhancing technologies for the internet, ii: Five years later," in Privacy Enhancing Technologies, R. Dingledine and P. Syverson, Eds., Berlin, Heidelberg: Springer Berlin Heidelberg, 2003, pp. 1–12, ISBN: 978-3-540-36467-2.
155. S. Garfinkel and H. R. Lipford, "Usable security: History, themes, and challenges," Synthesis Lectures on Information Security, Privacy, and Trust, vol. 5, no. 2, pp. 1–124, 2014.
156. S. Gaw, E. W. Felten, and P. Fernandez-Kelly, "Secrecy, flagging, and paranoia: Adoption criteria in encrypted email," in Proceedings of the SIGCHI Conference on Human Factors in Computing Systems, ser. CHI '06, Montréal, Québec, Canada: Association for Computing Machinery, 2006, pp. 591–600, ISBN: 1595933727. DOI: https://doi.org/10.1145/1124772.1124862. [Online]. Available: https://doi.org/10.1145/1124772.1124862.
157. G. Borradaile, K. Kretschmer, M. Gretes, and A. LeClerc, "The motivated can encrypt (even with pgp)," Proceedings on Privacy Enhancing Technologies, vol. 2021, no. 3, pp. 49–69, 2021.
158. C. Reuter, L. L. Iacono, and A. Benlian, A quarter century of usable security and privacy research: Transparency, tailorability, and the road ahead, 2022.
159. J. L. Dupree, R. Devries, D. M. Berry, and E. Lank, "Privacy personas: Clustering users via attitudes and behaviors toward security practices," in Proceedings of the 2016 CHI Conference on Human Factors in Computing Systems, ser. CHI '16, San Jose, California, USA: Association for Computing Machinery, 2016, pp. 5228–5239, ISBN: 9781450333627. DOI: https://doi.org/10.1145/2858036.2858214. [Online]. Available: https://doi.org/10.1145/2858036.2858214.
160. A. Morton and M. A. Sasse, "Desperately seeking assurances: Segmenting users by their information-seeking preferences," in 2014 Twelfth Annual International Conference on Privacy, Security and Trust, IEEE, Toronto, ON, Canada: IEEE, 2014, pp. 102–111. DOI: https://doi.org/10.1109/PST.2014.6890929.
161. O. R. Sanchez, I. Torre, Y. He, and B. P. Knijnenburg, "A recommendation approach for user privacy preferences in the fitness domain," User Modeling and User-Adapted Interaction, vol. 30, no. 3, pp. 513–565, 2020.

162. A. Acquisti, I. Adjerid, R. Balebako, L. Brandimarte, L. F. Cranor, S. Komanduri, P. G. Leon, N. Sadeh, F. Schaub, M. Sleeper, et al., "Nudges for privacy and security: Understanding and assisting users choices online," ACM Computing Surveys (CSUR), vol. 50, no. 3, pp. 1–41, 2017.
163. B. Liu, M. S. Andersen, F. Schaub, H. Almuhimedi, S. A. Zhang, N. Sadeh, Y. Agarwal, and A. Acquisti, "Follow my recommendations: A personalized privacy assistant for mobile app permissions," in Twelfth symposium on usable privacy and security (SOUPS 2016), Santa Clara, CA: USENIX Association, 2016, pp. 27–41.
164. B. Rashidi, C. Fung, and T. Vu, "Dude, ask the experts!: Android resource access permission recommendation with recdroid," in 2015 IFIP/IEEE international symposium on integrated network management (IM), Ottawa, ON, Canada: IEEE, 2015, pp. 296–304.
165. J. Pruitt and J. Grudin, "Personas: Practice and theory," in Proceedings of the 2003 Conference on Designing for User Experiences, ser. DUX '03, San Francisco, California: Association for Computing Machinery, 2003, pp. 1–15, ISBN: 1581137281. DOI: https://doi.org/10.1145/997078.997089. [Online]. Available: https://doi.org/10.1145/997078.997089.
166. J. Grudin and J. Pruitt, "Personas, participatory design and product development: An infrastructure for engagement," in Proc. PDC, vol. 2, 2202 N. 41st Street, Seattle, WA 98103: CPSR, 2002, pp. 144–152.
167. R. Sinha, "Persona development for information-rich domains," in CHI '03 Extended Abstracts on Human Factors in Computing Systems, ser. CHI EA '03, Ft. Lauderdale, Florida, USA: Association for Computing Machinery, 2003, pp. 830–831, ISBN: 1581136374. DOI: https://doi.org/10.1145/765891.766017. [Online]. Available: https://doi.org/10.1145/765891.766017.
168. J. (McGinn and N. Kotamraju, "Data-driven persona development," in Proceedings of the SIGCHI Conference on Human Factors in Computing Systems, ser. CHI '08, Florence, Italy: Association for Computing Machinery, 2008, pp. 1521–1524, ISBN: 9781605580111. DOI: https://doi.org/10.1145/1357054.1357292. [Online]. Available: https://doi.org/10.1145/1357054.1357292.
169. A. Cooper, "The inmates are running the asylum," in Software-Ergonomie '99: Design von Informationswelten. Wiesbaden: Vieweg+Teubner Verlag, 1999, pp. 17–17, ISBN: 978-3-322-99786-9. DOI: https://doi.org/10.1007/978-3-322-99786-9_1. [Online]. Available: https://doi.org/10.1007/978-3-322-99786-9_1.
170. P. Kumaraguru and L. F. Cranor, Privacy indexes: a survey of Westin's studies. Pennsylvania, USA: Carnegie Mellon University, School of Computer Science, Institute for Software Research, 2005.
171. T. Biselli, E. Steinbrink, F. Herbert, G. M. Schmidbauer-Wolf, and C. Reuter, "On the Challenges of Developing a Concise Questionnaire to Identify Privacy Personas," Proceedings on Privacy Enhancing Technologies (PoPETs), vol. 4, pp. 645–669, 2022. [Online]. Available: https://petsymposium.org/2022/files/papers/issue4/popets-2022-0126.pdf.
172. U. Holmström, "User-centered design of secure software," in Proceedings of Human Factors in Telecommunications, USA: Citeseer, 1999.
173. H. Ray, R. K. Flynn Wolf, and A. J. Aviv, "warn them or just block them?: Investigating privacy concerns among older and working age adults," Proceedings on Privacy Enhancing Technologies, vol. 2, pp. 27–47, 2021.

174. T. Biselli, L. Utz, and C. Reuter, "Supporting Informed Choices about Browser Cookies: The Impact of Personalised Cookie Banners," Proceedings on Privacy Enhancing Technologies (PoPETs), vol. 2024, pp. 171–191, 1 2024. DOI:https://doi.org/10.56553/popets-2024-0011. [Online]. Available: https://petsymposium.org/popets/2024/popets-2024-0011.pdf.
175. J. Lee, S. Rajtmajer, E. Srivatsavaya, and S. Wilson, "Digital inequality through the lens of self-disclosure," Proceedings on Privacy Enhancing Technologies, vol. 2021, no. 3, pp. 373–393, 2021.
176. N. Gerber, B. Reinheimer, and M. Volkamer, "Investigating people's privacy risk perception.," Proc. Priv. Enhancing Technol., vol. 2019, no. 3, pp. 267–288, 2019.
177. R. MacDonald and R. Smith, "Towards interface specification and design guidelines to raise user awareness of application security," computer, vol. 16, p. 24, 2004.
178. E. Spero and R. Biddle, "Out of sight, out of mind: Ui design and the inhibition of mental models of security," in New security paradigms workshop 2020, New York, NY, USA: Association for Computing Machinery, 2020, pp. 127–143.
179. A. R. Senarath and N. A. G. Arachchilage, "Understanding user privacy expectations: A software developers perspective," Telematics and Informatics, vol. 35, no. 7, pp. 1845–1862, 2018.
180. M. Namara, H. Sloan, and B. P. Knijnenburg, "The effectiveness of adaptation methods in improving user engagement and privacy protection on social network sites," Proceedings on Privacy Enhancing Technologies, vol. 2022, no. 1, pp. 629–648, 2022.
181. A. Mathur, J. Vitak, A. Narayanan, and M. Chetty, "Characterizing the use of {browser-based} blocking extensions to prevent online tracking," in Fourteenth symposium on usable privacy and security (SOUPS 2018), Santa Clara, CA: USENIX Association, 2018, pp. 103–116.
182. E. von Zezschwitz, S. Chen, and E. Stark, "" it builds trust with the customers"-exploring user perceptions of the padlock icon in browser ui," in 2022 IEEE Security and Privacy Workshops (SPW), San Francisco, CA, USA: IEEE, 2022, pp. 44–50.
183. M. Mehrnezhad, K. Coopamootoo, and E. Toreini, "How can and would people protect from online tracking?" Proceedings on Privacy Enhancing Technologies, vol. 1, pp. 105–125, 2022.
184. D. Smullen, Y. Yao, Y. Feng, N. Sadeh, A. Edelstein, and R. Weiss, "Managing potentially intrusive practices in the browser: A user-centered perspective," Proceedings on Privacy Enhancing Technologies, vol. 2021, no. 4, pp. 500–527, 2021.
185. F. M. Farke, D. G. Balash, M. Golla, M. Dürmuth, and A. J. Aviv, "Are privacy dashboards good for end users? evaluating user perceptions and reactions to google's my activity," in 30th USENIX Security Symposium (USENIX Security 21), virtual event: USENIX Association, 2021, pp. 483–500.
186. B. Weinshel, M. Wei, M. Mondal, E. Choi, S. Shan, C. Dolin, M. L. Mazurek, and B. Ur, "Oh, the places you've been! user reactions to longitudinal transparency about third-party web tracking and inferencing," in Proceedings of the 2019 ACM SIGSAC Conference on Computer and Communications Security, New York, NY, USA: Association for Computing Machinery, 2019, pp. 149–166.
187. K. Connelly, A. Khalil, and Y. Liu, "Do i do what i say?: Observed versus stated privacy preferences," Lecture Notes in Computer Science, vol. 4662, p. 620, 2007.
188. S. Kokolakis, "Privacy attitudes and privacy behaviour: A review of current research on the privacy paradox phenomenon," Computers & security, vol. 64, pp. 122–134, 2017.

189. R. E. Crossler and F. Bélanger, "Why would i use location-protective settings on my smartphone? motivating protective behaviors and the existence of the privacy knowledge–belief gap," Information Systems Research, vol. 30, no. 3, pp. 995–1006, 2019. DOI: https://doi.org/10.1287/isre.2019.0846. eprint: https://doi.org/10.1287/isre.2019.0846. [Online]. Available: https://doi.org/10.1287/isre.2019.0846.
190. P. K. Masur, D. Teutsch, and S. Trepte, "Entwicklung und validierung der online-privatheitskompetenzskala (oplis)," Diagnostica, vol. 63, no. 4, pp. 256–268, 2017.
191. N. K. Malhotra, S. S. Kim, and J. Agarwal, "Internet users' information privacy concerns (iuipc): The construct, the scale, and a causal model," Information systems research, vol. 15, no. 4, pp. 336–355, 2004.
192. L. Baruh, E. Secinti, and Z. Cemalcilar, "Online privacy concerns and privacy management: A meta-analytical review," Journal of Communication, vol. 67, no. 1, pp. 26–53, 2017.
193. H.-T. Chen and W. Chen, "Couldn't or wouldn't? the influence of privacy concerns and self-efficacy in privacy management on privacy protection," Cyberpsychology, Behavior, and Social Networking, vol. 18, no. 1, pp. 13–19, 2015.
194. A. R. Brough and K. D. Martin, "Critical roles of knowledge and motivation in privacy research," Current Opinion in Psychology, vol. 31, pp. 11–15, 2020, Privacy and Disclosure, Online and in Social Interactions, ISSN: 2352-250X. DOI: https://doi.org/10.1016/j.copsyc.2019.06.021. [Online]. Available: https://www.sciencedirect.com/science/article/pii/S2352250X19300648.
195. T. Groß, "Validity and reliability of the scale internet users' information privacy concern (iuipc)[extended version]," arXiv preprint arXiv:2011.11749, vol. abs/2011.11749, 2020.
196. S. Preibusch, "Guide to measuring privacy concern: Review of survey and observational instruments," International journal of human-computer studies, vol. 71, no. 12, pp. 1133–1143, 2013.
197. K. Kautz, "Investigating the design process: Participatory design in agile software development," Information Technology & People, vol. 24, no. 3, pp. 217–235, 2011.
198. E. Peer, L. Brandimarte, S. Samat, and A. Acquisti, "Beyond the turk: Alternative platforms for crowdsourcing behavioral research," Journal of Experimental Social Psychology, vol. 70, pp. 153–163, 2017, ISSN: 0022-1031. DOI: https://doi.org/10.1016/j.jesp.2017.01.006. [Online]. Available: https://www.sciencedirect.com/science/article/pii/S0022103116303201.
199. T. L. Adams, Y. Li, and H. Liu, "A replication of beyond the turk: Alternative platforms for crowdsourcing behavioral research sometimes preferable to student groups," AIS Transactions on Replication Research, vol. 6, p. 15, 1 2020, ISSN: 2473-3458. DOI: https://doi.org/10.17705/1atrr.00058. [Online]. Available: https://aisel.aisnet.org/trr/vol6/iss1/15.
200. K. Caine, "Local standards for sample size at chi," in Proceedings of the 2016 CHI Conference on Human Factors in Computing Systems, ser. CHI '16, San Jose, California, USA: Association for Computing Machinery, 2016, pp. 981–992, ISBN: 9781450333627. DOI: https://doi.org/10.1145/2858036.2858498. [Online]. Available: https://doi.org/10.1145/2858036.2858498.
201. B. Laugwitz, T. Held, and M. Schrepp, "Construction and evaluation of a user experience questionnaire," in HCI and Usability for Education and Work: 4th Symposium of the Workgroup Human-Computer Interaction and Usability Engineering of the Austrian

Computer Society, USAB 2008, Graz, Austria, November 20-21, 2008. Proceedings 4, vol. 5298, 2008, pp. 63–76, ISBN: 978-3-540-89349-3. DOI: https://doi.org/10.1007/978-3-540-89350-9_6.

202. M. Van Someren, Y. F. Barnard, and J. Sandberg, "The think aloud method: A practical approach to modelling cognitive," London: AcademicPress, vol. 11, pp. 29–41, 1994.
203. H. B. Mann and D. R. Whitney, "On a test of whether one of two random variables is stochastically larger than the other," The annals of mathematical statistics, vol. 18, no. 1, pp. 50–60, 1947.
204. T. Groß, "Toward valid and reliable privacy concern scales: The example of iuipc-8," in Human Factors in Privacy Research. Cham: Springer International Publishing, 2023, pp. 55–81, ISBN: 978-3-031-28643-8. DOI:https://doi.org/10.1007/978-3-031-28643-8_4. [Online]. Available: https://doi.org/10.1007/978-3-031-28643-8_4.
205. W. H. Hassan et al., "Current research on internet of things (iot) security: A survey," Computer networks, vol. 148, pp. 283–294, 2019.
206. Z.-K. Zhang, M. C. Y. Cho, C.-W. Wang, C.-W. Hsu, C.-K. Chen, and S. Shieh, "Iot security: Ongoing challenges and research opportunities," in 2014 IEEE 7th International Conference on Service-Oriented Computing and Applications, Matsue, Japan: IEEE, 2014, pp. 230–234. DOI:https://doi.org/10.1109/SOCA.2014.58.
207. O. Alrawi, C. Lever, M. Antonakakis, and F. Monrose, "SoK: security evaluation of home-based iot deployments," in IEEE Symposium on Security and Privacy (SP), San Francisco, CA, USA: IEEE, 2019, pp. 1362–1380. DOI: https://doi.org/10.1109/SP.2019.00013.
208. C. Kolias, G. Kambourakis, A. Stavrou, and J. Voas, "Ddos in the iot: Mirai and other botnets," Computer, vol. 50, no. 7, pp. 80–84, 2017. DOI: https://doi.org/10.1109/MC.2017.201.
209. S. Zheng, N. Apthorpe, M. Chetty, and N. Feamster, "User perceptions of smart home iot privacy," Proceedings of the ACM on human-computer interaction, vol. 2, no. CSCW, pp. 1–20, 2018.
210. E. Zeng, S. Mare, and F. Roesner, "End user security and privacy concerns with smart homes," in Proceedings of the Thirteenth USENIX Conference on Usable Privacy and Security, ser. SOUPS '17, vol. 220, Santa Clara, CA, USA: USENIX Association, 2017, pp. 65–80.
211. B. Kuang, A. Fu, S. Yu, G. Yang, M. Su, and Y. Zhang, "Esdra: An efficient and secure distributed remote attestation scheme for iot swarms," IEEE Internet of Things Journal, vol. 6, no. 5, pp. 8372–8383, 2019.
212. E. Dushku, M. M. Rabbani, M. Conti, L. V. Mancini, and S. Ranise, "Sara: Secure asynchronous remote attestation for iot systems," IEEE Transactions on Information Forensics and Security, vol. 15, pp. 3123–3136, 2020.
213. B. Kuang, A. Fu, L. Zhou, W. Susilo, and Y. Zhang, "Do-ra: Data-oriented runtime attestation for iot devices," Computers & Security, vol. 97, p. 101 945, 2020.
214. X. Carpent, N. Rattanavipanon, and G. Tsudik, "Remote attestation of iot devices via smarm: Shuffled measurements against roving malware," in 2018 IEEE international symposium on hardware oriented security and trust (HOST), Washington, DC, USA: IEEE, 2018, pp. 9–16.
215. M. Conti, E. Dushku, and L. V. Mancini, "Radis: Remote attestation of distributed iot services," in 2019 Sixth International Conference on Software Defined Systems (SDS), Rome, Italy: IEEE, 2019, pp. 25–32.

216. M. Ammar, B. Crispo, and G. Tsudik, "Simple: A remote attestation approach for resource-constrained iot devices," in 2020 ACM/IEEE 11th International Conference on Cyber-Physical Systems (ICCPS), Sydney, NSW, Australia: IEEE, 2020, pp. 247–258. DOI: https://doi.org/10.1109/ICCPS48487.2020.00036.
217. I. D. O. Nunes, K. Eldefrawy, N. Rattanavipanon, M. Steiner, and G. Tsudik, "Vrased: A verified hardware/software co-design for remote attestation," in Proceedings of the 28th USENIX Conference on Security Symposium, ser. SEC'19, Santa Clara, CA, USA: USENIX Association, 2019, pp. 1429–1446.
218. T. Abera, N. Asokan, L. Davi, J.-E. Ekberg, T. Nyman, A. Paverd, A.-R. Sadeghi, and G. Tsudik, "C-flat: Control-flow attestation for embedded systems software," in Proceedings of the 2016 ACM SIGSAC Conference on Computer and Communications Security, ser. CCS '16, Vienna, Austria: Association for Computing Machinery, 2016, pp. 743–754, ISBN: 9781450341394. DOI: https://doi.org/10.1145/2976749.2978358. [Online]. Available: https://doi.org/10.1145/2976749.2978358.
219. T. Abera, R. Bahmani, F. Brasser, A. Ibrahim, A.-R. Sadeghi, and M. Schunter, "Diat: Data integrity attestation for resilient collaboration of autonomous systems.," in NDSS, San Diego, CA, USA: Network and Distributed Systems Security (NDSS), 2019, pp. 1–15.
220. I. D. O. Nunes, S. Jakkamsetti, and G. Tsudik, "Dialed: Data integrity attestation for low-end embedded devices," in 2021 58th ACM/IEEE Design Automation Conference (DAC), San Francisco, CA, USA: IEEE, 2021, pp. 313–318.
221. R. Roemer, E. Buchanan, H. Shacham, and S. Savage, "Return-oriented programming: Systems, languages, and applications," ACM Transactions on Information and System Security (TISSEC), vol. 15, no. 1, pp. 1–34, 2012.
222. H. Hu, S. Shinde, S. Adrian, Z. L. Chua, P. Saxena, and Z. Liang, "Data-oriented programming: On the expressiveness of non-control data attacks," in 2016 IEEE Symposium on Security and Privacy (SP), IEEE, San Jose, CA, USA: IEEE, 2016, pp. 969–986.
223. M. A. Sasse and I. Flechais, "Usable security: Why do we need it? how do we get it?" In Security and Usability: Designing secure systems that people can use, L. Cranor and S. Garfinkel, Eds., Sebastopol, US: O'Reilly, 2005, pp. 13–30.
224. S. Furnell and N. Clarke, "Power to the people? the evolving recognition of human aspects of security," computers & security, vol. 31, no. 8, pp. 983–988, 2012.
225. OWASP, Internet of things (iot) top 10 2018, 2018. [Online]. Available: https://owasp.org/www-pdf-archive/OWASP-IoT-Top-10-2018-final.pdf.
226. B. Nyhan and J. Reifler, "When corrections fail: The persistence of political misperceptions," Political Behavior, vol. 32, no. 2, pp. 303–330, 2010.
227. S. Surminski, C. Niesler, S. Linsner, L. Davi, and C. Reuter, "SCAtt-man: Side-Channel-Based Remote Attestation for Embedded Devices that Users Understand," in Proceedings of the Thirteenth ACM Conference on Data and Application Security and Privacy, ser. CODASPY '23, New York, NY, USA: Association for Computing Machinery, 2023, pp. 225–236, ISBN: 9798400700675. DOI: https://doi.org/10.1145/3577923.3583652. [Online]. Available: https://doi.org/10.1145/3577923.3583652.
228. L. Babun, Z. B. Celik, P. McDaniel, and A. S. Uluagac, "Real-time analysis of privacy-(un) aware iot applications," Proceedings on Privacy Enhancing Technologies, vol. 2021, no. 1, pp. 145–166, 2021.

229. S. Mare, F. Roesner, and T. Kohno, "Smart devices in airbnbs: Considering privacy and security for both guests and hosts.," Proc. Priv. Enhancing Technol., vol. 2020, no. 2, pp. 436–458, 2020.
230. M. N. Aman, M. H. Basheer, S. Dash, J. W. Wong, J. Xu, H. W. Lim, and B. Sikdar, "Hatt: Hybrid remote attestation for the internet of things with high availability," IEEE Internet of Things Journal, vol. 7, no. 8, pp. 7220–7233, 2020.
231. T. Abera, N. Asokan, L. Davi, F. Koushanfar, A. Paverd, A.-R. Sadeghi, and G. Tsudik, "Invited: Things, trouble, trust: On building trust in iot systems," in Proceedings of the 53rd Annual Design Automation Conference, Austin, TX, USA: ACM, IEEE, 2016, pp. 1–6. DOI: https://doi.org/10.1145/2897937.2905020.
232. C. Kil, E. C. Sezer, A. M. Azab, P. Ning, and X. Zhang, "Remote attestation to dynamic system properties: Towards providing complete system integrity evidence," in 2009 IEEE/IFIP International Conference on Dependable Systems & Networks, Lisbon, Portugal: IEEE, 2009, pp. 115–124. DOI: https://doi.org/10.1109/DSN.2009.5270348.
233. B. Parno, J. M. McCune, and A. Perrig, "Bootstrapping trust in commodity computers," in 2010 IEEE Symposium on Security and Privacy, Oakland, CA, USA: IEEE, 2010, pp. 414–429. DOI: https://doi.org/10.1109/SP.2010.32.
234. V. Costan and S. Devadas, "Intel sgx explained," Cryptology ePrint Archive, vol. Paper 2016, no. 086, 2016. [Online]. Available: https://eprint.iacr.org/2016/086.
235. A. ARM, "Arm security technology: Building a secure system using trustzone technology," ARM Limited, white paper, 2009.
236. S. Surminski, C. Niesler, F. Brasser, L. Davi, and A.-R. Sadeghi, "Realswatt: Remote software-based attestation for embedded devices under realtime constraints," in 2021 ACM SIGSAC Conference on Computer and Communications Security (CCS), ser. CCS '21, Virtual Event, Republic of Korea: ACM, 2021, pp. 2890–2905. DOI: https://doi.org/10.1145/3460120.3484788.
237. A. Seshadri, A. Perrig, L. van Doorn, and P. K. Khosla, "Swatt: Software-based attestation for embedded devices," in IEEE Symposium on Security and Privacy (SP), Berkeley, CA, USA: IEEE, 2004, pp. 272–282. DOI: https://doi.org/10.1109/SECPRI.2004.1301329.
238. P. Koeberl, S. Schulz, A.-R. Sadeghi, and V. Varadharajan, "Trustlite: A security architecture for tiny embedded devices," in Proceedings of the Ninth European Conference on Computer Systems, ser. EuroSys '14, Amsterdam, Netherlands: ACM, 2014. DOI: https://doi.org/10.1145/2592798.2592824.
239. F. Brasser, B. El Mahjoub, A.-R. Sadeghi, C. Wachsmann, and P. Koeberl, "Tytan: Tiny trust anchor for tiny devices," in Proceedings of the 52nd annual Design Automation Conference, San Francisco, CA, USA: ACM, IEEE, 2015, pp. 1–6. DOI: https://doi.org/10.1145/2744769.2744922.
240. M. Ambrosin, M. Conti, R. Lazzeretti, M. M. Rabbani, and S. Ranise, "Collective remote attestation at the internet of things scale: State-of-the-art and future challenges," IEEE Communications Surveys & Tutorials, vol. 22, no. 4, pp. 2447–2461, 2020.
241. K. Pfeffer, A. Mai, A. Dabrowski, M. Gusenbauer, P. Schindler, E. Weippl, M. Franz, and K. Krombholz, "On the usability of authenticity checks for hardware security tokens," in 30th USENIX Security Symposium (USENIX Security 21), virtual event: USENIX Association, 2021, pp. 37–54. [Online]. Available: https://www.usenix.org/conference/usenixsecurity21/presentation/pfeffer.

242. Y. Seralathan, T. T. Oh, S. Jadhav, J. Myers, J. P. Jeong, Y. H. Kim, and J. N. Kim, "Iot security vulnerability: A case study of a web camera," in 2018 20th International Conference on Advanced Communication Technology (ICACT), Chuncheon, Korea (South): IEEE, 2018, pp. 172–177. DOI: https://doi.org/10.23919/ICACT.2018.8323686.
243. C. Yan, X. Ji, K. Wang, Q. Jiang, Z. Jin, and W. Xu, "A survey on voice assistant security: Attacks and countermeasures," ACM Computing Surveys, vol. 55, no. 4, pp. 1–36, 2022.
244. E. Lafontaine, A. Sabir, and A. Das, "Understanding peoples attitude and concerns towards adopting iot devices," in Extended Abstracts of the 2021 CHI Conference on Human Factors in Computing Systems, ser. CHI EA '21, Yokohama, Japan: Association for Computing Machinery, 2021, ISBN: 9781450380959. DOI: https://doi.org/10.1145/3411763.3451633. [Online]. Available: https://doi.org/10.1145/3411763.3451633.
245. A. Zubiaga, R. Procter, and C. Maple, "A longitudinal analysis of the public perception of the opportunities and challenges of the internet of things," PloS one, vol. 13, no. 12, e0209472, 2018.
246. L. Nemec Zlatolas, N. Feher, and M. Hölbl, "Security perception of iot devices in smart homes," Journal of Cybersecurity and Privacy, vol. 2, no. 1, pp. 65–73, 2022.
247. C. Park, Y. Kim, and M. Jeong, "Influencing factors on risk perception of iot-based home energy management services," Telematics and Informatics, vol. 35, no. 8, pp. 2355–2365, 2018, ISSN: 0736-5853. DOI: https://doi.org/10.1016/j.tele.2018.10.005. [Online]. Available: https://www.sciencedirect.com/science/article/pii/S0736585318303939.
248. M. Asplund and S. Nadjm-Tehrani, "Attitudes and perceptions of iot security in critical societal services," IEEE Access, vol. 4, pp. 2130–2138, 2016. DOI: https://doi.org/10.1109/ACCESS.2016.2560919.
249. A. AlHogail, "Improving iot technology adoption through improving consumer trust," Technologies, vol. 6, no. 3, p. 64, 2018.
250. E. Korneeva, N. Olinder, and W. Strielkowski, "Consumer attitudes to the smart home technologies and the internet of things (iot)," Energies, vol. 14, no. 23, p. 7913, 2021.
251. M. N. Alraja, M. M. J. Farooque, and B. Khashab, "The effect of security, privacy, familiarity, and trust on users attitudes toward the use of the iot-based healthcare: The mediation role of risk perception," Ieee Access, vol. 7, pp. 111 341–111 354, 2019.
252. A. J. van Deursen, A. van der Zeeuw, P. de Boer, G. Jansen, and T. van Rompay, "Digital inequalities in the internet of things: Differences in attitudes, material access, skills, and usage," Information, Communication & Society, vol. 24, no. 2, pp. 258–276, 2021.
253. P. Oser, S. Feger, P. W. Woźniak, J. Karolus, D. Spagnuelo, A. Gupta, S. Lüders, A. Schmidt, and F. Kargl, "Safer: Development and evaluation of an iot device risk assessment framework in a multinational organization," Proc. ACM Interact. Mob. Wearable Ubiquitous Technol., vol. 4, no. 3, 2020. DOI: https://doi.org/10.1145/3414173. [Online]. Available: https://doi.org/10.1145/3414173.
254. A. Remesh, D. Muralidharan, N. Raj, J. Gopika, and P. K. Binu, "Intrusion detection system for iot devices," in 2020 International Conference on Electronics and Sustainable Communication Systems (ICESC), Coimbatore, India: IEEE, 2020, pp. 826–830. DOI: https://doi.org/10.1109/ICESC48915.2020.9155999.
255. P. Emami-Naeini, J. Dheenadhayalan, Y. Agarwal, and L. F. Cranor, "Which privacy and security attributes most impact consumers' risk perception and willingness to purchase iot devices?" In 2021 IEEE Symposium on Security and Privacy (SP), IEEE, San Francisco, CA, USA: IEEE, 2021, pp. 519–536.

256. C. D. McDermott, J. P. Isaacs, and A. V. Petrovski, "Evaluating awareness and perception of botnet activity within consumer internet-of-things (iot) networks," in Informatics, MDPI, vol. 6, Basel, Switzerland: MDPI, 2019, p. 8.
257. N. Abdi, K. M. Ramokapane, and J. M. Such, "More than smart speakers: Security and privacy perceptions of smart home personal assistants," in Fifteenth Symposium on Usable Privacy and Security (SOUPS 2019), Santa Clara, CA: USENIX Association, 2019, pp. 451–466. [Online]. Available: https://www.usenix.org/conference/soups2019/presentation/abdi.
258. N. Gerber, B. Reinheimer, and M. Volkamer, "Home sweet home? investigating users' awareness of smart home privacy threats," in Proceedings of An Interactive Workshop on the Human aspects of Smarthome Security and Privacy (WSSP), Baltimore, MD, USA: USENIX Association, 2018, p. 2018.
259. M. Williams, J. R. C. Nurse, and S. Creese, "Privacy is the boring bit: User perceptions and behaviour in the internet-of-things," in 2017 15th Annual Conference on Privacy, Security and Trust (PST), Calgary, AB, Canada: IEEE, 2017, pp. 181–18 109. DOI: https://doi.org/10.1109/PST.2017.00029.
260. M. Saeidi, M. Calvert, A. W. Au, A. Sarma, and R. B. Bobba, "If this context then that concern: Exploring users' concerns with ifttt applets," Proceedings on Privacy Enhancing Technologies, vol. 2022, no. 1, pp. 166–186, 2021.
261. A. Alshehri, J. Spielman, A. Prasad, and C. Yue, "Exploring the privacy concerns of bystanders in smart homes from the perspectives of both owners and bystanders," Proceedings on Privacy Enhancing Technologies, vol. 3, pp. 99–119, 2022.
262. J. Abbott, J. Dev, D. Kim, S. Gopavaram, M. Iyer, S. Sadam, S. Mare, T. Ringenberg, V. Andalibi, and L. J. Camp, "Privacy lessons learnt from deploying an iot ecosystem in the home," in Proceedings of the 2022 European Symposium on Usable Security, ser. EuroUSEC '22, Karlsruhe, Germany: Association for Computing Machinery, 2022, pp. 98–110, ISBN: 9781450397001. DOI: https://doi.org/10.1145/3549015.3554205. [Online]. Available: https://doi.org/10.1145/3549015.3554205.
263. J. S. Edu, J. M. Such, and G. Suarez-Tangil, "Smart home personal assistants: A security and privacy review," ACM Computing Surveys (CSUR), vol. 53, no. 6, pp. 1–36, 2020.
264. H. Aldowah, S. Ul Rehman, and I. Umar, "Trust in iot systems: A vision on the current issues, challenges, and recommended solutions," in Advances on Smart and Soft Computing, F. Saeed, T. Al-Hadhrami, F. Mohammed, and E. Mohammed, Eds., Singapore: Springer Singapore, 2021, pp. 329–339, ISBN: 978-981-15-6048-4.
265. P. Menard and G. J. Bott, "Analyzing iot users' mobile device privacy concerns: Extracting privacy permissions using a disclosure experiment," Computers & Security, vol. 95, p. 101 856, 2020, ISSN: 0167-4048. DOI: https://doi.org/10.1016/j.cose.2020.101856. [Online]. Available: https://www.sciencedirect.com/science/article/pii/S0167404820301280.
266. R. Mahmoud, T. Yousuf, F. Aloul, and I. Zualkernan, "Internet of things (iot) security: Current status, challenges and prospective measures," in 2015 10th International Conference for Internet Technology and Secured Transactions (ICITST), London, UK, 2015: IEEE, 2015, pp. 336–341. DOI: https://doi.org/10.1109/ICITST.2015.7412116.
267. Business Wire. "Strategy analytics: Global smart speaker sales cross 150 million units for 2020 following robust q4 demand, 2023 Business Wire, Inc. (2021), [Online]. Available: https://www.businesswire.com/news/home/20210303005852/en/Strategy-Analytics-Global-Smart-Speaker-Sales-Cross-150-Million-Units-for-2020-Following-Robust-Q4-Demand.

268. D. Bawden and L. Robinson, "Information overload: An overview," in Oxford Encyclopedia of Political Decision Making, Oxford: Oxford University Press, 2020. DOI: https://doi.org/10.1093/acrefore/9780190228637.013.1360. [Online]. Available: https://openaccess.city.ac.uk/id/eprint/23544/.
269. S. Moss. "Acquiescence bias," Sicotests 2016, (2016), [Online]. Available: https://www.sicotests.com/newpsyarticle/Acquiescence-bias.
270. M. Schrepp, A. Hinderks, and J. Thomaschewski, "Design and evaluation of a short version of the user experience questionnaire (ueq-s)," International Journal of Interactive Multimedia and Artificial Intelligence, vol. 4, no. 6, pp. 103–108, 2017. DOI: https://doi.org/10.9781/ijimai.2017.09.001.
271. C. Reuter, M.-A. Kaufhold, S. Schmid, T. Spielhofer, and A. S. Hahne, "The Impact of Risk Cultures: Citizens' Perception of Social Media Use in Emergencies across Europe," Technological Forecasting and Social Change (TFSC), vol. 148, no. 119724, pp. 1–17, 2019. DOI: https://doi.org/10.1016/j.techfore.2019.119724. [Online]. Available: http://www.peasec.de/paper/2019/2019_ReuterKaufholdSchmidSpielhoferHahne_TheImpactofRiskCultures_TFSC.pdf.
272. H. Ayatollahi, P. Bath, and S. Goodacre, "Accessibility versus confidentiality of information in the emergency department," Emergency Medicine Journal, vol. 26, no. 12, pp. 857–860, 2009.
273. G. T. Duncan, T. B. Jabine, and V. A. de Wolf, Private lives and public policies: Confidentiality and accessibility of government statistics. National Academy Press, 1993.
274. E. Yanakieva, M. Youssef, A. H. Rezae, and A. Bieniusa, "On the impossibility of confidentiality, integrity and accessibility in highly-available file systems," in Networked Systems: 9th International Conference, NETYS 2021, Virtual Event, May 19–21, 2021, Proceedings, Springer, 2021, pp. 3–18.
275. X. de Carné de Carnavalet and P. C. van Oorschot, "A survey and analysis of TLS interception mechanisms and motivations," CoRR, vol. abs/2010.16388, 2020. arXiv: https://arxiv.org/abs/2010.16388. [Online]. Available: https://arxiv.org/abs/2010.16388.
276. J. Ning, X. Huang, G. S. Poh, S. Xu, J.-C. Loh, J. Weng, and R. H. Deng, "Pine: Enabling privacy-preserving deep packet inspection on tls with rule-hiding and fast connection establishment," in Computer Security–ESORICS 2020: 25th European Symposium on Research in Computer Security, ESORICS 2020, Guildford, UK, September 14–18, 2020, Proceedings, Part I 25, Springer, 2020, pp. 3–22.
277. T. Ringenberg and L. Kisselburgh, "A way forward: What we know (or not) about {csam} & privacy," 2022.
278. J. Prokos, N. Fendley, M. Green, R. Schuster, E. Tromer, T. M. Jois, and Y. Cao, "Squint hard enough: Attacking perceptual hashing with adversarial machine learning," in USENIX Security Symposium, USENIX Association, 2023.
279. S. Jain, A. Cretu, and Y. de Montjoye, "Adversarial detection avoidance attacks: Evaluating the robustness of perceptual hashing-based client-side scanning," in USENIX Security Symposium, USENIX Association, 2022, pp. 2317–2334.
280. J. Sherry, C. Lan, R. A. Popa, and S. Ratnasamy, "Blindbox: Deep packet inspection over encrypted traffic," in Proceedings of the 2015 ACM Conference on Special Interest Group on Data Communication, SIGCOMM 2015, London, United Kingdom, August 17-21, 2015, S. Uhlig, O. Maennel, B. Karp, and J. Padhye, Eds., ACM, 2015, pp. 213–226. DOI: https://doi.org/10.1145/2785956.2787502. [Online]. Available: https://doi.org/10.1145/2785956.2787502.

281. C. Zhang, Z. DeStefano, A. Arun, J. Bonneau, P. Grubbs, and M. Walfish, "Zombie: Middleboxes that don't snoop," IACR Cryptol. ePrint Arch., p. 1022, 2023.
282. P. Grubbs, A. Arun, Y. Zhang, J. Bonneau, and M. Walfish, "Zero-knowledge middleboxes," in USENIX Security Symposium, USENIX Association, 2022, pp. 4255–4272.
283. D. Naylor, K. Schomp, M. Varvello, I. Leontiadis, J. Blackburn, D. R. López, K. Papagiannaki, P. R. Rodrıguez, and P. Steenkiste, "Multi-context TLS (mctls): Enabling secure in-network functionality in TLS," in SIGCOMM, ACM, 2015, pp. 199–212.
284. M. Fischlin, Stealth key exchange and confined access to the record protocol data in tls 1.3, Cryptology ePrint Archive, Paper 2023/651, to appear at CCS 2023, 2023. [Online]. Available: https://eprint.iacr.org/2023/651.
285. A. Lerner, E. Zeng, and F. Roesner, "Confidante: Usable encrypted email: A case study with lawyers and journalists," in 2017 IEEE European Symposium on Security and Privacy (EuroS&P), Institute of Electrical and Electronics Engineers Inc., 2017, pp. 385–400, ISBN: 9781509057610. DOI: https://doi.org/10.1109/EuroSP.2017.41.
286. M. T. de Oliveira, L. H. A. Reis, R. C. Carrano, F. L. Seixas, D. C. M. Saade, C. V. Albuquerque, N. C. Fernandes, S. D. Olabarriaga, D. S. V. Medeiros, and D. M. F. Mattos, "Towards a blockchain-based secure electronic medical record for healthcare applications," in ICC 2019 - 2019 IEEE International Conference on Communications (ICC), 2019, pp. 1–6. DOI: https://doi.org/10.1109/ICC.2019.8761307.
287. M. Li, C. Lal, M. Conti, and D. Hu, "Lechain: A blockchain-based lawful evidence management scheme for digital forensics," Future Generation Computer Systems, vol. 115, pp. 406–420, 2021, ISSN: 0167739X. DOI: https://doi.org/10.1016/j.future.2020.09.038.
288. H. Zheng, L. You, and G. Hu, "A novel insurance claim blockchain scheme based on zero-knowledge proof technology," Computer Communications, vol. 195, pp. 207–216, 2022, ISSN: 1873703X. DOI: https://doi.org/10.1016/j.comcom.2022.08.007.
289. A. S. Yadav, V. Charles, D. K. Pandey, and S. Gupta, "Blockchain-based secure privacy-preserving vehicle accident and insurance registration," Expert Systems with Applications, vol. 230, 2023, ISSN: 09574174. DOI: https://doi.org/10.1016/j.eswa.2023.120651.
290. K. Thapliyal and A. Pathak, "Quantum e-commerce: A comparative study of possible protocols for online shopping and other tasks related to e-commerce," Quantum Information Processing, vol. 18, 6 2019, ISSN: 15731332. DOI: https://doi.org/10.1007/s11128-019-2309-3.
291. Y.-H. Chou, F.-J. Lin, and G.-J. Zeng, "An efficient novel online shopping mechanism based on quantum communication," Electronic Commerce Research, vol. 14, pp. 349–367, 3 2014, ISSN: 15729362. DOI: https://doi.org/10.1007/s10660-014-9143-6.
292. W. Huang, Y.-H. Yang, and H.-Y. Jia, "Cryptanalysis and improvement of a quantum communication-based online shopping mechanism," Quantum Information Processing, vol. 14, pp. 2211–2225, 6 2015, ISSN: 15700755. DOI: https://doi.org/10.1007/s11128-015-0958-4.
293. The Guardian, Eu lawyers say plan to scan private messages for child abuse may be unlawful, accessed August 21, 2023, 2023. [Online]. Available: https://www.theguardian.com/world/2023/may/08/eu-lawyers-plan-to-scan-private-messages-child-abuse-may-be-unlawful-chat-controls-regulation.
294. X. de Carné de Carnavalet and M. Mannan, "Killed by proxy: Analyzing client-end tls interception software," in Network and Distributed System Security Symposium, Internet Society, 2017. DOI: https://doi.org/10.14722/ndss.2016.23374.

295. M. Green, R. Droms, R. Housley, P. Turner, and S. Fenter, "Data Center use of Static Diffie-Hellman in TLS 1.3," Internet Engineering Task Force, Internet-Draft draft-green-tls-static-dh-in-tls13-01, 2017, Work in Progress, 15 pp. [Online]. Available: https://datatracker.ietf.org/doc/html/draft-green-tls-static-dh-in-tls13-01.
296. H. Lee, Z. Smith, J. Lim, G. Choi, S. Chun, T. Chung, and T. T. Kwon, "Matls: How to make TLS middlebox-aware?" In NDSS, The Internet Society, 2019.
297. X. de Carné de Carnavalet and M. Mannan, "Killed by proxy: Analyzing client-end TLS interception software," in NDSS, The Internet Society, 2016.
298. G. Ateniese, D. H. Chou, B. de Medeiros, and G. Tsudik, "Sanitizable signatures," in Computer Security – ESORICS 2005, S. d. C. di Vimercati, P. Syverson, and D. Gollmann, Eds., Berlin, Heidelberg: Springer Berlin Heidelberg, 2005, pp. 159–177, ISBN: 978-3-540-31981-8.
299. V. Fehr and M. Fischlin, "Sanitizable signcryption: Sanitization over encrypted data (full version)," IACR Cryptol. ePrint Arch., vol. 2015, p. 765, 2015. [Online]. Available: https://api.semanticscholar.org/CorpusID:16085003.
300. G. Fuchsbauer, R. Gay, L. Kowalczyk, and C. Orlandi, "Access control encryption for equality, comparison, and more," in Public-Key Cryptography – PKC 2017, S. Fehr, Ed., Berlin, Heidelberg: Springer Berlin Heidelberg, 2017, pp. 88–118, ISBN: 978-3-662-54388-7.
301. I. Damgård, H. Haagh, and C. Orlandi, "Access control encryption: Enforcing information flow with cryptography," in Theory of Cryptography, M. Hirt and A. Smith, Eds., Berlin, Heidelberg: Springer Berlin Heidelberg, 2016, pp. 547–576, ISBN: 978-3-662-53644-5.
302. S. Kim and D. J. Wu, Access control encryption for general policies from standard assumptions, Cryptology ePrint Archive, Paper 2017/467, https://eprint.iacr.org/2017/467, 2017. [Online]. Available: https://eprint.iacr.org/2017/467.
303. X. Wang and S. S. M. Chow, "Cross-domain access control encryption: Arbitrary-policy, constant-size, efficient," in 2021 IEEE Symposium on Security and Privacy (SP), 2021, pp. 748–761. DOI: https://doi.org/10.1109/SP40001.2021.00023.
304. L. Assouline and B. Minaud, "Weighted oblivious ram, with applications to searchable symmetric encryption," in Advances in Cryptology – EUROCRYPT 2023: 42nd Annual International Conference on the Theory and Applications of Cryptographic Techniques, Lyon, France, April 23-27, 2023, Proceedings, Part I, Lyon, France: Springer-Verlag, 2023, pp. 426–455, ISBN: 978-3-031-30544-3. DOI: https://doi.org/10.1007/978-3-031-30545-0_15. [Online]. Available: https://doi.org/10.1007/978-3-031-30545-0_15.
305. E. M. Kornaropoulos, N. Moyer, C. Papamanthou, and A. Psomas, "Leakage inversion: Towards quantifying privacy in searchable encryption," in Proceedings of the 2022 ACM SIGSAC Conference on Computer and Communications Security, ser. CCS '22, Los Angeles, CA, USA: Association for Computing Machinery, 2022, pp. 1829–1842, ISBN: 9781450394505. DOI:https://doi.org/10.1145/3548606.3560593. [Online]. Available: https://doi.org/10.1145/3548606.3560593.
306. K. S. Kim, M. Kim, D. Lee, J. H. Park, and W.-H. Kim, "Forward secure dynamic searchable symmetric encryption with efficient updates," in Proceedings of the 2017 ACM SIGSAC Conference on Computer and Communications Security, ser. CCS '17, Dallas, Texas, USA: Association for Computing Machinery, 2017, pp. 1449–1463, ISBN: 9781450349468. DOI: https://doi.org/10.1145/3133956.3133970. [Online]. Available: https://doi.org/10.1145/3133956.3133970.

307. R. Bost, B. Minaud, and O. Ohrimenko, "Forward and backward private searchable encryption from constrained cryptographic primitives," in Proceedings of the 2017 ACM SIGSAC Conference on Computer and Communications Security, ser. CCS '17, Dallas, Texas, USA: Association for Computing Machinery, 2017, pp. 1465–1482, ISBN: 9781450349468. DOI: https://doi.org/10.1145/3133956.3133980. [Online]. Available: https://doi.org/10.1145/3133956.3133980.
308. S. Kamara and T. Moataz, "Boolean searchable symmetric encryption with worst-case sub-linear complexity," in Advances in Cryptology – EUROCRYPT 2017, J.-S. Coron and J. B. Nielsen, Eds., Cham: Springer International Publishing, 2017, pp. 94–124, ISBN: 978-3-319-56617-7.
309. S.-F. Sun, X. Yuan, J. K. Liu, R. Steinfeld, A. Sakzad, V. Vo, and S. Nepal, "Practical backward-secure searchable encryption from symmetric puncturable encryption," in Proceedings of the 2018 ACM SIGSAC Conference on Computer and Communications Security, ser. CCS '18, Toronto, Canada: Association for Computing Machinery, 2018, pp. 763–780, ISBN: 9781450356930.
310. D. X. Song, D. Wagner, and A. Perrig, "Practical techniques for searches on encrypted data," in Proceeding 2000 IEEE Symposium on Security and Privacy. S&P 2000, 2000, pp. 44–55. DOI: https://doi.org/10.1109/SECPRI.2000.848445.
311. I. Demertzis, D. Papadopoulos, and C. Papamanthou, "Searchable encryption with optimal locality: Achieving sublogarithmic read efficiency," in Advances in Cryptology – CRYPTO 2018: 38th Annual International Cryptology Conference, Santa Barbara, CA, USA, August 19–23, 2018, Proceedings, Part I, Santa Barbara, CA, USA: Springer-Verlag, 2018, pp. 371–406, ISBN: 978-3-319-96883-4. DOI: https://doi.org/10.1007/978-3-319-96884-1_13. [Online]. Available: https://doi.org/10.1007/978-3-319-96884-1_13.
312. S. Lai, S. Patranabis, A. Sakzad, J. K. Liu, D. Mukhopadhyay, R. Steinfeld, S.-F. Sun, D. Liu, and C. Zuo, "Result pattern hiding searchable encryption for conjunctive queries," in Proceedings of the 2018 ACM SIGSAC Conference on Computer and Communications Security, ser. CCS '18, Toronto, Canada: Association for Computing Machinery, 2018, pp. 745–762, ISBN: 9781450356930. DOI:https://doi.org/10.1145/3243734.3243753. [Online]. Available: https://doi.org/10.1145/3243734.3243753.
313. S. Patel, G. Persiano, and K. Yeo, "Lower bounds for encrypted multi-maps and searchable encryption in the leakage cell probe model," in Advances in Cryptology – CRYPTO 2020, D. Micciancio and T. Ristenpart, Eds., Cham: Springer International Publishing, 2020, pp. 433–463, ISBN: 978-3-030-56784-2.
314. A. Bossuat, R. Bost, P.-A. Fouque, B. Minaud, and M. Reichle, "Sse and ssd: Page-efficient searchable symmetric encryption," in Advances in Cryptology – CRYPTO 2021, T. Malkin and C. Peikert, Eds., Cham: Springer International Publishing, 2021, pp. 157–184, ISBN: 978-3-030-84252-9.
315. G. Asharov, G. Segev, and I. Shahaf, "Tight tradeoffs in searchable symmetric encryption," in Advances in Cryptology – CRYPTO 2018, H. Shacham and A. Boldyreva, Eds., Cham: Springer International Publishing, 2018, pp. 407–436, ISBN: 978-3-319-96884-1.
316. B. Minaud and M. Reichle, "Dynamic local searchable symmetric encryption," CoRR, vol. abs/2201.05006, 2022. arXiv: https://arxiv.org/abs/2201.05006. [Online]. Available: https://arxiv.org/abs/2201.05006.
317. J. Ghareh Chamani, D. Papadopoulos, C. Papamanthou, and R. Jalili, "New constructions for forward and backward private symmetric searchable encryption," in Proceed-

ings of the 2018 ACM SIGSAC Conference on Computer and Communications Security, ser. CCS '18, Toronto, Canada: Association for Computing Machinery, 2018, pp. 1038–1055, ISBN: 9781450356930. DOI: https://doi.org/10.1145/3243734.3243833. [Online]. Available: https://doi.org/10.1145/3243734.3243833.

318. R. Dingledine, N. Mathewson, and P. F. Syverson, "Tor: The second-generation onion router," in USENIX Security Symposium, USENIX, 2004, pp. 303–320.
319. N. Andriamilanto and T. Allard, "Brfast: A tool to select browser fingerprinting attributes for web authentication according to a usability-security trade-off," in Companion Proceedings of the Web Conference 2021, ser. WWW '21, Ljubljana, Slovenia: Association for Computing Machinery, 2021, pp. 701–704, ISBN: 9781450383134. DOI: https://doi.org/10.1145/3442442.3458610. [Online]. Available: https://doi.org/10.1145/3442442.3458610.
320. N. Ben-Asher, J. Meyer, Y. Parmet, S. Moeller, and R. Englert, "An experimental microworld for evaluating the tradeoffs between usability and security," in Symposium on Usable Privacy and Security (SOUPS), 2010.
321. M. A. Sasse and M. Smith, "The security-usability tradeoff myth [guest editors' introduction]," IEEE Security & Privacy, vol. 14, no. 5, pp. 11–13, 2016. DOI: https://doi.org/10.1109/MSP.2016.102.
322. M. A. Sasse, M. Smith, C. Herley, H. Lipford, and K. Vaniea, "Debunking security-usability tradeoff myths," IEEE Security & Privacy, vol. 14, no. 5, pp. 33–39, 2016. DOI: https://doi.org/10.1109/MSP.2016.110.
323. F. D. Nocera and G. Tempestini, "Getting rid of the usability/security trade-off: A behavioral approach," Journal of Cybersecurity and Privacy, vol. 2, pp. 245–256, 2 2022. DOI: https://doi.org/10.3390/jcp2020013.
324. M. Minami, K. Suzaki, and T. Okumura, "Security considered harmful a case study of tradeoff between security and usability," in 2011 IEEE Consumer Communications and Networking Conference (CCNC), 2011, pp. 523–524. DOI: https://doi.org/10.1109/CCNC.2011.5766529.
325. A. Whitten and J. D. Tygar, "Why johnny can't encrypt: A usability evaluation of pgp 5.0.," in USENIX security symposium, vol. 348, 1999, pp. 169–184.
326. S. Ruoti, J. Andersen, D. Zappala, and K. E. Seamons, "Why johnny still, still can't encrypt: Evaluating the usability of a modern PGP client," CoRR, vol. abs/1510.08555, 2015. arXiv: https://arxiv.org/abs/1510.08555. [Online]. Available: http://arxiv.org/abs/1510.08555.
327. S. Sheng, L. Broderick, C. A. Koranda, and J. J. Hyland, "Why johnny still can't encrypt: Evaluating the usability of email encryption software," in Symposium on usable privacy and security, ACM, 2006, pp. 3–4.
328. A. Reuter, A. Abdelmaksoud, K. Boudaoud, and M. Winckler, "Usability of end-to-end encryption in e-mail communication," Frontiers in Big Data, vol. 4, 2021, ISSN: 2624909X. DOI: https://doi.org/10.3389/fdata.2021.568284.
329. W. Bai, D. Kim, M. Namara, Y. Qian, P. G. Kelley, and M. L. Mazurek, "Balancing security and usability in encrypted email," IEEE Internet Computing, vol. 21, no. 3, pp. 30–38, 2017. DOI: https://doi.org/10.1109/MIC.2017.57.
330. C. Tiefenau, E. von Zezschwitz, M. Häring, K. Krombholz, and M. Smith, "A usability evaluation of let's encrypt and certbot: Usable security done right," in Proceedings of the 2019 ACM SIGSAC Conference on Computer and Communications Security,

Association for Computing Machinery, 2019, pp. 1971–1988, ISBN: 9781450367479. DOI: https://doi.org/10.1145/3319535.3363220.

331. G. Borradaile, K. Kretschmer, M. Gretes, and A. LeClerc, "The motivated can encrypt (even with pgp)," Proceedings on Privacy Enhancing Technologies, vol. 2021, pp. 49–69, 3 2021. DOI: https://doi.org/10.2478/popets-2021-0037.
332. E. A. Qahtani, Y. Javed, and M. Shehab, "User perceptions of gmails confidential mode," Proceedings on Privacy Enhancing Technologies, vol. 2022, pp. 187–206, 1 2022. DOI: https://doi.org/10.2478/popets-2022-0010.
333. S. Landau, "Control use of data to protect privacy," Science, vol. 347, no. 6221, pp. 504–506, 2015.
334. J. Colnago, L. Cranor, and A. Acquisti, "Is there a reverse privacy paradox? an exploratory analysis of gaps between privacy perspectives and privacy-seeking behaviors," Proceedings on Privacy Enhancing Technologies, vol. 2023, pp. 455–476, 1 2023. DOI: https://doi.org/10.56553/popets-2023-0027.
335. H. Ray, F. Wolf, R. Kuber, and A. J. Aviv, ""warn them" or "just block them"?: Investigating privacy concerns among older and working age adults," Proceedings on Privacy Enhancing Technologies, vol. 2021, pp. 27–47, 2 2021. DOI: https://doi.org/10.2478/popets-2021-0016.
336. N. Gerber, B. Berens, and M. Volkamer, "Investigating people's privacy risk perception," Proceedings on Privacy Enhancing Technologies, vol. 2019, pp. 267–288, 2019. DOI: https://doi.org/10.2478/popets-2019-0047.
337. E. L. Olmsted-Hawala, E. D. Murphy, S. Hawala, and K. T. Ashenfelter, "Think-aloud protocols: A comparison of three think-aloud protocols for use in testing data-dissemination web sites for usability," in Proceedings of the SIGCHI Conference on Human Factors in Computing Systems, ser. CHI '10, Atlanta, Georgia, USA: Association for Computing Machinery, 2010, pp. 2381–2390, ISBN: 9781605589299. DOI: https://doi.org/10.1145/1753326.1753685. [Online]. Available: https://doi.org/10.1145/1753326.1753685.
338. J. Olson, C. McAllister, L. D. Grinnell, K. Gehrke Walters, and F. Appunn, "Applying constant comparative method with multiple investigators and inter-coder reliability," 2016.
339. K. Caine, "Local standards for sample size at chi," in Proceedings of the 2016 CHI Conference on Human Factors in Computing Systems, ser. CHI '16, San Jose, California, USA: Association for Computing Machinery, 2016, pp. 981–992, ISBN: 9781450333627. DOI: https://doi.org/10.1145/2858036.2858498. [Online]. Available: https://doi.org/10.1145/2858036.2858498.
340. C. Fritz, P. Morris, and J. Richler, "Effect size estimates: Current use, calculations, and interpretation," Journal of experimental psychology. General, vol. 141, pp. 2–18, 2011. DOI: https://doi.org/10.1037/a0024338.
341. J. Cohen, Statistical power analysis for the behavioral sciences. 1988, ISBN: 9780203771587. DOI: https://doi.org/10.4324/9780203771587.
342. P. Stachwitz and J. F. Debatin, "Digitalisierung im gesundheitswesen: Heute und in zukunft," Bundesgesundheitsblatt-Gesundheitsforschung-Gesundheitsschutz, vol. 66, no. 2, pp. 105–113, 2023.
343. F. Herbert, S. Becker, L. Schaewitz, J. Hielscher, M. Kowalewski, A. Sasse, Y. Acar, and M. Dürmuth, "A world full of privacy and security (mis) conceptions? findings of a

representative survey in 12 countries," in Proceedings of the 2023 CHI Conference on Human Factors in Computing Systems, 2023, pp. 1–23.

344. C. Prince, N. Omrani, A. Maalaoui, M. Dabic, and S. Kraus, "Are we living in surveillance societies and is privacy an illusion? an empirical study on privacy literacy and privacy concerns," IEEE Transactions on Engineering Management, 2021.
345. M. Harbach, A. De Luca, N. Malkin, and S. Egelman, "Keep on lockin'in the free world: A multi-national comparison of smartphone locking," in Proceedings of the 2016 chi conference on human factors in computing systems, 2016, pp. 4823–4827.
346. C. Utz, S. Becker, T. Schnitzler, F. M. Farke, F. Herbert, L. Schaewitz, M. Degeling, and M. Dürmuth, "Apps against the spread: Privacy implications and user acceptance of covid-19-related smartphone apps on three continents," in Proceedings of the 2021 chi conference on human factors in computing systems, 2021, pp. 1–22.
347. A. Ilhan and K. J. Fietkiewicz, "Data privacy-related behavior and concerns of activity tracking technology users from germany and the usa," Aslib Journal of Information Management, vol. 73, no. 2, pp. 180–200, 2021.
348. L. E. Pleger, K. Guirguis, and A. Mertes, "Making public concerns tangible: An empirical study of german and uk citizens perception of data protection and data security," Computers in Human Behavior, vol. 122, p. 106 830, 2021.
349. C. Sindermann, H. S. Schmitt, F. Kargl, C. Herbert, and C. Montag, "Online privacy literacy and online privacy behavior – the role of crystallized intelligence and personality," International Journal of Human–Computer Interaction, vol. 37, no. 15, pp. 1455–1466, 2021. DOI: https://doi.org/10.1080/10447318.2021.1894799. eprint: https://doi.org/10.1080/10447318.2021.1894799. [Online]. Available: https://doi.org/10.1080/10447318.2021.1894799.
350. A.-M. Ortloff, S. Zimmerman, D. Elsweiler, and N. Henze, "The effect of nudges and boosts on browsing privacy in a naturalistic environment," in Proceedings of the 2021 Conference on Human Information Interaction and Retrieval, ser. CHIIR '21, Canberra ACT, Australia: Association for Computing Machinery, 2021, pp. 63–73, ISBN: 9781450380553. DOI: https://doi.org/10.1145/3406522.3446014. [Online]. Available: https://doi.org/10.1145/3406522.3446014.
351. E. Ntoutsi, K. Stefanidis, K. Nørvåg, and H.-P. Kriegel, "Fast group recommendations by applying user clustering," in International conference on conceptual modeling, Springer, 2012, pp. 126–140.
352. T. Franke, C. Attig, and D. Wessel, "A personal resource for technology interaction: Development and validation of the affinity for technology interaction (ati) scale," International Journal of Human–Computer Interaction, vol. 35, no. 6, pp. 456–467, 2019.
353. M. Gutfleisch, M. Schöps, J. Hielscher, M. Cheney, S. Sayin, N. Schuhmacher, A. Mohamad, and M. A. Sasse, "Caring about iot-security–an interview study in the healthcare sector," in Proceedings of the 2022 European Symposium on Usable Security, ser. EuroUSEC '22, New York, NY, USA: Association for Computing Machinery, 2022, pp. 202–215.
354. M. Kunze. "Turning google smart speakers into wiretaps for $100k." (2022), [Online]. Available: https://downrightnifty.me/blog/2022/12/26/hacking-google-home.html.
355. J. M. Haney, Y. Acar, and S. Furman, " "it's the company, the government, you and i": User perceptions of responsibility for smart home privacy and security," in USENIX Security Symposium, USENIX Association, 2021, pp. 411–428,

ISBN: 978-1-939133-24-3. [Online]. Available: https://www.usenix.org/conference/usenixsecurity21/presentation/haney.

356. P. E. McKnight and J. Najab, "Mann-whitney u test," in The Corsini Encyclopedia of Psychology, I. Weiner and W. Craighead, Eds., Weinheim, Deutschland: John Wiley & Sons, Ltd, 2010, pp. 1–1. DOI: https://doi.org/10.1002/9780470479216.corpsy0524.
357. R. Wash and E. Rader, "Too much knowledge? security beliefs and protective behaviors among united states internet users," in Eleventh Symposium On Usable Privacy and Security (SOUPS 2015), 2015, pp. 309–325.
358. B. Ur, J. Bees, S. M. Segreti, L. Bauer, N. Christin, and L. F. Cranor, "Do users' perceptions of password security match reality?" In Proceedings of the 2016 CHI conference on human factors in computing systems, 2016, pp. 3748–3760.
359. S. Fahl, M. Harbach, T. Muders, M. Smith, and U. Sander, "Helping johnny 2.0 to encrypt his facebook conversations," p. 216, 2012.
360. M. Green and M. Smith, "Developers are not the enemy!: The need for usable security apis," IEEE Security & Privacy, vol. 14, no. 5, pp. 40–46, 2016. DOI: https://doi.org/10.1109/MSP.2016.111.
361. G. Goos, J. Hartmanis, J. Van, L. E. Board, D. Hutchison, T. Kanade, J. Kittler, J. M. Kleinberg, F. Mattern, E. Zurich, J. C. Mitchell, M. Naor, O. Nierstrasz, B. Steffen, M. Sudan, D. Terzopoulos, D. Tygar, M. Y. Vardi, and G. Weikum, "Lncs 3679 - computer security esorics 2005," 2005.
362. Ghostery GmbH: Ghostery, Accessed on 2024-12-04. [Online]. Available: https://www.ghostery.com/.
363. Disconnect, Accessed on 2024-12-04. [Online]. Available: https://disconnect.me/disconnect.
364. Duck Duck Go Inc.: Duck Duck Go Essentials, Accessed on 2024-12-04. [Online]. Available: https://duckduckgo.com/.
365. G. Maone, Noscript, Accessed on 2024-12-04. [Online]. Available: https://noscript.net/.
366. R. Hill, Umatrix, Accessed on 2024-12-04. [Online]. Available: https://github.com/gorhill/uMatrix.

The manufacturer's authorised representative in the EU is Springer Nature Customer Service Centre GmbH, Europaplatz 3, 69115 Heidelberg, Germany. If you have any concerns regarding our products, please contact ProductSafety@springernature.com

Printed and bound by CPI Group (UK) Ltd, Croydon, CR0 4YY
30/09/2026
02225566-0004